Diversity of Oceanic Life: An Evaluative Review

SIGNIFICANT ISSUES SERIES papers are written for and published by the Center for Strategic and International Studies.
Series Editors: David M. Abshire
                 Douglas M. Johnston, Jr.
Director of Studies: Erik R. Peterson
Director of Publications: Nancy B. Eddy
Managing Editor: Roberta L. Howard
Associate Editor: Susan Sojourner

The Center for Strategic and International Studies (CSIS), founded in 1962, is an independent, tax-exempt, public policy research institution based in Washington, D.C.

The mission of the Center is to advance the understanding of emerging world issues in the areas of international economics, politics, security, and business. It does so by providing a strategic perspective to decision makers that is integrative in nature, international in scope, anticipatory in its timing, and bipartisan in its approach. The Center's commitment is to serve the common interests and values of the United States and other countries around the world that support representative government and the rule of law.

The Honolulu-based **Pacific Forum**, which merged with CSIS in 1989, retains its autonomy as a nonprofit policy institute, researching and analyzing political, economic, and security interests and trends in the Asia-Pacific region. It is the hub of a network of 20 research institutes around the Pacific Rim, drawing on Asian perspectives and disseminating its various projects' findings and recommendations to opinion leaders and decision makers throughout the region. Its activities are fully coordinated with the CSIS Asian Studies Program.

The Center for Strategic and International Studies
1800 K Street, N.W., Suite 400
Washington, D.C. 20006
Telephone: (202) 887-0200
Telex: 7108229583
Fax: (202) 775-3199
Cable Address: CENSTRAT

Volume XIV, Number 12
Significant Issues Series

Diversity of Oceanic Life
An Evaluative Review

edited by Melvin N.A. Peterson

foreword by Frederick Seitz

Ocean Policy Institute
Pacific Forum/CSIS
Honolulu

The Center for Strategic
and International Studies
Washington, D.C.

Library of Congress Cataloging-in-Publication Data

Diversity of oceanic life : an evaluative review / edited by Melvin N. A.
    Peterson : contributions by Martin V. Angel . . . [et al.].
        p.   cm. — (Significant issues series, ISSN 0736-7136 : v. 14,
    no. 12)
        Includes bibliographical references.
        ISBN 0-89206-206-1
        1. Marine biology—Congresses. 2. Biological diversity conservation—
    Congresses. 3. Biological diversity conservation—Government policy—
    Congresses. 4. Marine resources conservation—Congresses. I. Peterson,
    Melvin N. A. II. Angel, Martin Vivian. III. Series.
    QH91.D58 1992
    333.95′216—dc20   92-39818
                                                                                      CIP

The Center for Strategic and International Studies, as an independent
research institution, does not take specific public policy positions.
Accordingly, all views, positions, and conclusions expressed in
this publication should be understood to be solely those of the individual
authors. Neither does this report reflect the position of the parent
institutions of the participants.

Graphic art for chapter 2 created by Magellan Mapmakers—Jess Zimmerman.

© 1992

The Center for Strategic and International Studies
Washington, D.C.

All Rights Reserved

Printed on Recycled Paper

Contents

Acknowledgments	vii
Foreword *by Frederick Seitz*	ix
Study Participants	xi
About the Contributors	xiii
Executive Summary	xv

Part I. Evaluative Summary

1. Diversity of Life: Its Meaning and Significance in Oceanic Realms *by Melvin N. A. Peterson*   1
- Introduction   1
- Discussion   6
- Priorities   12
- Capabilities and Goals   21

Part II. Background Papers

2. Managing Biodiversity in the Oceans *by Martin V. Angel*   23
- What is Biodiversity?   23
- Why are the Oceans Special?   25
- Global Variability in Time and Space   28
- The Species Concept   32
- How the Diversity Concept Developed   34
- Biodiversity in the Oceans   38
- Biodiversity and Environmental Management   49
- Deep Ocean Conservation   53
- Conclusions   56

3. Oceanic Species Diversity *by John A. McGowan*   60
- Recommendations   62

4. Biodiversity: Human Impacts through Fisheries and Transportation *by Makoto Omori, Christopher P. Norman, and Hiroshi Yamakawa*   63
- Fishing Activity   64
- Inadvertent Transport of Microorganisms   65
- Enhancement Fisheries   67
- Discussion   70

5. **Mariculture on Coastal China and Biodiversity**
   by C. K. Tseng 76
   The Historical Perspective 76
   Seaweed Mariculture 78
   Mollusk Mariculture 85
   Shrimp Mariculture 89
   Fish Mariculture 91
   Overall Development of Mariculture on
      Coastal China 93
   Mariculture and Biodiversity 93

6. **Microbial Diversity** *by Rita R. Colwell
   and Russell Hill* 100

About the Ocean Policy Institute 107

Advisory Council of the Ocean Policy Institute,
   Pacific Forum/CSIS 109

Acknowledgments

The Ocean Policy Institute of the Pacific Forum/CSIS gratefully acknowledges a grant for institutional support from the state of Hawaii and the implementing contract from the University of Hawaii. We also gratefully acknowledge the generous and visionary grants from Cynthia Leising, Edward W. Scripps, Jr., and William H. Scripps that made this work possible under the Edward W. Scripps II Program in Conservation of Ocean Resources.

Foreword

This is the inaugural study of the new Ocean Policy Institute of the Pacific Forum/CSIS. It is appropriate that this first study should be a policy analysis that applies at several different levels of policy.

The study assesses the quality of our scientific knowledge of oceanic life and the level of effort to advance the field; the study also makes thoughtful recommendations for future work. Thus, it is clearly a commentary directed toward ocean science policy and toward biological science policy.

At a broader level, practically every scientific or technical talent known to humanity can be exercised in regard to oceanic realms. In this sense, then, the analysis relates to all science and technology policy.

In assessing the quality of our knowledge base that should serve to help undergird many societal decisions and policy formulations with respect to oceans and environment, the study also makes welcome comment concerning the factual basis for this level of decision. Particularly for the coastal nations, these are decisions that can affect their national life. The wide range of interests and experience to be found in our Advisory Council to this new Institute is illustrative of the range of national decisions that can involve the oceans.

Finally, the oceans lie between the lands and are inherently a subject of international concern. This study has significantly engaged an international group of experts. Their recommendations are based in their broad view of a field in which international cooperation is both possible and desirable.

Valuable and informative, with its view for the future, this analysis should be carefully read by all who find either appeal or importance in the oceans and their life.

Frederick Seitz
Chairman, Advisory Council, Ocean Policy Institute,
Former President, National Academy of Sciences, and
Former President, Rockefeller University

Study Participants
Marine Biodiversity Project of the Ocean Policy Institute
Pacific Forum/CSIS

Isabella Abbott, professor of botany, University of Hawaii (Honolulu, Hawaii)

Martin V. Angel, chairman, Biology Department, Institute of Oceanographic Sciences, Deacon Laboratory, (United Kingdom)

John Bardach, interim director, Environmental and Policy Institute, The East-West Center (Honolulu, Hawaii)

Rita R. Colwell, president, Maryland Biotechnology Institute, and former member, National Science Board (College Park, Maryland)

Sylvia Earle, former chief scientist, National Oceanic and Atmospheric Administration (Washington, D.C.)

Glenn Flittner, director, Office of Research and Environmental Information, National Marine Fisheries Service, National Oceanic and Atmospheric Administration (Washington, D.C.)

David Greenfield, ichthyologist, and dean, Graduate Division, University of Hawaii (Honolulu, Hawaii)

Elizabeth A. Kay, professor of zoology, University of Hawaii (Honolulu, Hawaii)

Phil Lobel, associate scientist, Woods Hole Oceanographic Institute (Woods Hole, Massachusetts)

James Maragos, chief scientist, The Nature Conservancy, Pacific Region (Honolulu, Hawaii)

John McGowan, professor of oceanography, Scripps Institution of Oceanography of the University of California (San Diego, California)

Makoto Omori, professor, Department of Aquatic Biosciences, Tokyo University of Fisheries (Tokyo, Japan)

Brian Rothschild, professor, Chesapeake Biological Laboratory, University of Maryland (Solomons, Maryland)

C.K. Tseng, director emeritus, Institute of Oceanology, Academia Sinica (Qingdao, People's Republic of China)

Elizabeth Venrick, professor of marine biology, Scripps Institution of Oceanography of the University of California (San Diego, California)

About the Contributors

Martin V. Angel studied zoology at Cambridge and Bristol Universities, United Kingdom (UK). After participating in the International Indian Ocean Expedition for 18 months in 1963–1964, he decided to take up a career in biological oceanography, joining what is now the Institute of Oceanographic Sciences, Deacon Laboratory (UK) in 1965. He specialized in planktonic ecology and now heads the biology department. He advises the World Wildlife Fund (UK) on oceanic conservation problems and is currently vice president of the British Ecological Society.

Rita R. Colwell is president of the Maryland Biotechnology Institute and professor of microbiology at the University of Maryland at College Park. She obtained her Ph.D. in marine microbiology from the University of Washington. Dr. Colwell has focused much of her research on marine biotechnology and the molecular genetics of marine and estuarine bacteria, and also on the microbiology of the Chesapeake Bay, coastal recreational waters, and the deep ocean.

Russell Hill is a faculty member and colleague of Rita Colwell at the University of Maryland.

John A. McGowan is professor of oceanography in the graduate department of Scripps Institution of Oceanography of the University of California at San Diego. Professor McGowan has spent his professional life in teaching and research in oceanography. Generally recognized as one of the world's foremost experts in planktonic life of the open ocean, he has made many fundamental contributions to what might be called the oceanography of life.

Christopher P. Norman is a postdoctoral fellow at Tokyo University of Fisheries, specializing in marine ecology.

Makoto Omori is a professor in the department of aquatic biosciences at Tokyo University of Fisheries and an adjunct professor in the hydrographic department of the Maritime Safety Agency. He obtained his Ph.D. from Hokkaido University, Japan, in 1966. His specialty is biological oceanography. During his career he has been a research assistant at Woods Hole Oceanographic Institute; assistant professor at the Ocean

Research Institute, University of Tokyo; associate biological oceanographer at Scripps Institution of Oceanography; program specialist, Division of Marine Sciences, United Nations Educational, Scientific and Cultural Organization (UNESCO); and associate professor at Tokyo University of Fisheries.

Melvin N. A. Peterson, director of the Ocean Policy Institute, Pacific Forum/CSIS in Honolulu, Hawaii, organized and edited this evaluative study. He is former chief scientist of the National Oceanic and Atmospheric Administration within the U.S. Department of Commerce. After receiving his Ph.D. from Harvard, he spent almost three decades on the faculty of Scripps Institution of Oceanography where he also served as chief scientist, program director, and principal investigator of the Deep Sea Drilling Project, with its scientific drilling vessel "Glomar Challenger."

C. K. Tseng has had a long and distinguished career that embraces many roles, from research scientist and world expert in algae, to research leader and longtime director within the institute structure of the Chinese Academy of Sciences, to respected and admired statesman for science, both within China and internationally. As director emeritus of the Institute of Oceanology in Qingdao, Dr. Tseng's view reaches back many decades from his earlier educational experiences in both the People's Republic of China, where he received his bachelor of science degree from Amoy University (1931), and the United States, where he received his doctor of science degree from the University of Michigan (1942), through the development of modern mariculture in China. He has been honored in many ways, including an honorary doctorate from Ohio State University (1987), the China National Science Award (1956 and again in 1987), and the Scientist with Extraordinary Contribution award, Shandong, China (1992).

Hiroshi Yamakawa is an associate professor in the department of aquatic biosciences, Tokyo University of Fisheries, specializing in mariculture and aquaculture.

Executive Summary

- In any concerns about global change, it is biologic change that should most concern people because it is life and its processes that truly condition the planet and provide the trophic infrastructure that translates solar energy to the food on which humans depend. Biological systems also play a major role in maintaining environmental quality by cleaning up, breaking down, and detoxifying natural and many artificial wastes. Oceanic life may be one of the most sensitive indicators of integrated change.
- The 70 percent of the earth's surface that is oceanic holds a commanding position with regard to climate, photosynthesis, and evaporation/precipitation patterns, hence on our food production, on availability of living resources, and even on the oxygen in the air we breath.
- Increasing fisheries, pollution, mariculture, intentional and unintentional introductions or modification of species, and other human impacts are creating unprecedented potential threats to the normal biological interactions of the oceans.
- Broadly viewed, proper mariculture may offer greater good, despite its impacts, by providing food, reducing the need for traditional fishing, and increasing incentives to avoid oceanic and coastal water pollution. New biotechnologies offer major advances in food-producing efficiencies and rearability of mariculture stocks, in new biochemical products, and in dramatically new culture technologies. These new prospects depend on and can help protect marine life.
- There are major issues and opportunities. Yet societies now give pathetically little attention to the study of oceanic life.
- Scientific studies of oceanic life—including the most basic capabilities for describing its variability, its geographic distribution, and its natural relationships—are seriously deficient.
- Just as world consciousness has been driven to the need to be concerned about the basic integrity of oceanic ecosystems, the capacity of trained people to meet the increasing demands for knowledge is diminishing at an alarming rate.

It is clear that the field will not be able to meet society's expectations if this trend is not reversed. Knowledge of oceanic diversity is meager. Senior researchers and experts in describing and identifying marine organisms and studying their natural relationships and evolutionary development are aging; young people are not entering the field in sufficient numbers, nor are there the professional incentives to do so. We are near the point of losing knowledge, capability, and even existing information in a field in which many years of experience go into making fine experts. These circumstances must be set right.

- Recent and wonderful advances in population genetics and genetic manipulation, in cellular and molecular biology, and in biochemistry are welcome; but for the oceans we really are still in need of the basic knowledge, description, and understanding of oceanic life and its diversity. Advanced technical fields depend on basic biologic descriptions and species identifications, just as they will also help to refine them.
- Fishing is an ancient and worthy practice and has more recently been the subject of extensive oceanographic research, but systematic scientific study and description of oceanic life have received a very late start compared to similar studies on land. The oceans are vast, remote, deep, and difficult to study.
- Large gaps in knowledge exist. Some major aspects of marine ecosystems do not seem to fit at all well with precepts and theories derived from land; underlying facts must be obtained and more suitable concepts developed for the oceans.
- A clear and broad need exists to review, organize, and disseminate the most current information and classifications of marine organisms, their natural relationships, and the development of their speciation and distribution.
- The study of oceanic life is ready for a *recapitulation of knowledge* and a *renaissance,* driven by needs, opportunities, and new technologies.
- Study and professional development to expand description of biodiversity and biogeography and life's natural relationships in the oceans are equally essential with these same issues on land, and they must be improved if global ecological conditions are to be scientifically predictable and managed wisely.

- Renewed attention must be paid in all areas and institutions of biological sciences and public policy to ensure that ocean life sciences are supported at sufficiently greater levels to permit existing research expertise to be applied to global-scale problems and to assure the continuance and refinement of this expertise.
- A shift in priority toward basic data gathering, description, and dissemination must be undertaken for the ocean biological sciences to develop better understanding and appreciation and a sound policy for managing ocean life's abundance and diversity in light of its related social, environmental, and economic implications.
- Society in general and the national and international funding agencies and other sources specifically must redress the imbalance in research allocation priorities to give greater importance to studies of oceanic life and its many manifestations of diversity, interdependence, and wholeness.
- Species and within-species knowledge of biodiversity and other oceanographic knowledge must be integrated into ecosystem models and regional management regimes in systematic programs to ensure that valid information is available and used for problem-solving and policy-making.
- We must develop a comprehensive, computer-based global listing and information management system for descriptions of oceanic life—including its various manifestations of diversity, geography, and natural relationships—to capture, organize, and make available existing information, to provide efficiency for work in the future, and to serve as a teaching, research, and environmental management capability. Current information management systems and capabilities should be examined, refined, and adapted to these needs.
- We must develop a plan for collections as an essential part of this specimen-based science and its data needs.
- Major national and international programs that are now being aggressively planned to address understanding and predicting earth's climate system from space, land, and sea offer enormous opportunities for study of oceanic life. These programs can make possible symbiotic and cost-effective utilization of the data flow and the oceanic components of the

technical and logistic infrastructure. Planning should embrace these opportunities. Systematic availability of synoptic oceanic data in operational time scales for biological studies of the oceans is assumed, but should be reviewed as part of overall planning.

- Team approaches will be increasingly important in these multidisciplinary and transdisciplinary areas, in which specialists in oceanic taxonomy, systematics, and natural history will occupy basic and integrating roles, nationally and internationally, as appropriate. Scientists from small or developing countries can then participate in the broad enterprise.
- These integrating programs and studies are rightly to be done in the international, cooperative, and traditionally scientific mode and will facilitate the availability of an accepted factual base that will help prevent otherwise poorly formulated, generalized environmental concerns and associated strife.
- As ideological and military issues continue, as we hope, to become less important, enlarged or revised agendas of security for and among nations will increasingly confer greater importance on ecological and environmental security.
- Simply inventing the answers will not suffice.
- Whole ocean stewardship must be an ultimate objective.

1
Diversity of Life: Its Meaning and Significance in Oceanic Realms
Melvin N. A. Peterson

Introduction

Oceanic life is fully as important to planetary well-being and human prosperity and health as life on land. The oceans have had a unique planetary presence and permanence, and oceanic realms display a complexity all their own. These waters have been the cradle of life on earth and most singularly distinguish the earth from the other planets in the solar system.

Scientists increasingly recognize the coupling of the ocean and the atmosphere in earth's climate system. Nations now undertake major programs to study and understand that system, to distinguish natural from human-induced variations, and to predict possible future change.

Yet, in any concerns about global change, it is *biologic change* that should most concern people because it is life and its processes that truly condition the planet and provide the trophic infrastructure that translates solar energy to the food on which humans depend. Biological systems also play a major role in maintaining environmental quality by cleaning up, breaking down, and detoxifying natural and many artificial wastes. Oceanic life may be one of the most sensitive indicators of integrated change. The concept of biodiversity attempts to embrace the full richness of life on earth; the need for meaningful protection of this richness attracts increasing attention.

Realizing the enormous significance of the oceans to these broad issues and their associated future needs and opportunities, the Ocean Policy Institute of the Pacific Forum/CSIS initiated as one of its inaugural projects a review of the adequacy of our knowledge of oceanic life and its diversity. Underlying this decision was the perception, within the Institute and its advisory council, that the status of the field and efforts to advance it are not proportional to the field's importance from scientific, economic, environmental, and associated policy viewpoints.

In the summer of 1992, the Institute asked a group of the foremost experts in marine biology to consider the status of our understanding of oceanic life and its many dimensions of diversity. In addition, special review papers were prepared for these deliberations and are published here as essential components of this report. Each of these papers addresses a specific part of the broad subject.

Dr. Martin V. Angel, of the Institute of Oceanographic Sciences Deacon Laboratory in the United Kingdom, addresses the broad subject of managing biodiversity in the oceans. His presentation describes the existing and developing concept of biodiversity and relates it to the oceans and our knowledge of the oceans and their life. He also examines the origins of variability and diversity with respect to the changing geometry, processes, and conditions of the oceans and the different time scales of events and processes. These time scales range from the very longest—the full evolutionary time scale or much of the planetary time scale at billions of years—to those associated with significant development or change in the shapes and sizes of oceanic basins and the positions of the continents at tens of millions of years, to those of the waxing and waning of continental glaciers at tens of thousands of years, to the much more ephemeral events such as development of hydrothermal vents and their unique organic communities that may endure only for decades, to the familiar annual, tidal, and daily cycles. Issues of stability and change, of isolation and communication, and of duration and rate of change all bear on the uniqueness of oceanic life, its wide range of adaptations, and the validity of scientific understanding of it. Dr. Angel then turns to the issues and difficulties of translating this understanding into practices and principles for conceivable management, emphasizing the profound differences between land and sea and the important inadequacies of applying ecological principles derived from the study of life and its diversity on land to the interpretation and management of life and ecosystems of the oceans.

Oceanic realms contain great domains that can be characterized by factors such as the amount of light, the circulation of their waters, the richness of nutrients, and their relative permanence. Dr. John A. McGowan of the Scripps Institution of Oceanography, University of California, draws on a lifetime of experience in the study of several of these great domains in the open ocean. He discusses the state of our knowledge and future prospects and needs concerning the plankton communities, the myriad small plants and animals that live in and effectively drift along with the great wind-driven circulations of the oceans. Some of these broad motions circulate and substantially recirculate surface waters, such as in the anticyclonic gyres that occupy the vast central regions of the Pacific to the north and south of the tropics; in these instances nutrients are largely used up by the organisms, creating a general nutrient economy of scarcity. These circulations also afford an internal continuity to carry the organisms in a sufficiently closed loop to permit their stable presence and reproduction. Other open ocean circulations can exchange their waters more readily and draw in or mix with more nutrient-rich waters; the Equatorial Current system, with components flowing both east and west, is such a system and may offer a more complex closed loop for the organisms' drift and reproductive continuity. The southward flowing and cold California current, between North America and the central North Pacific gyre, draws up and entrains even more nutrient-rich waters and is an economy of abundance; except for local and small currents, it is dominantly a one-way route and offers little in the way of closed loop recirculation for organisms' reproductive continuity. These great circulations are driven by the latitudinal wind belts (for example, prevailing westerlies and trades) and surely have had long-term generic permanence as major components of the ocean-atmosphere system that redistributes low-latitude solar heat on a rotating earth. Addressing the basic life forms—the primary photosynthetic producers and small animals that feed on them and form the base of the food web in the open ocean—Dr. McGowan challenges the fundamental understanding of the development and maintenance of species diversity in these realms and points out that theories from land seem inadequate to explain the diversity patterns of the oceans. At issue, of course, is the robustness of the basic fabric of these marine ecosystems, important because of the large proportion of the earth they occupy and influence and because of the extent of human impacts that are now possible.

Dr. Makoto Omori, with colleagues from the Tokyo University of Fisheries, addresses such impacts, particularly as they may result from fisheries and aquaculture and through maritime transportation. Roughly 100 million tons of top predators—predators such as tuna that feed high in the food chain—are taken in marine fisheries each year. In some situations there is little doubt that fishing pressure has depleted stocks and led to significant if not permanent change. In other situations natural variability seems to dominate. Dr. Omori and his colleagues discuss values and environmental cautions surrounding the culture of seafood resources in terms of both constrained culture techniques and unconstrained release of juvenile stages into the natural environment. Intentional and unintentional releases of non-native organisms present possible risks, just as do introductions of weeds and pests on land. Jeopardy of natural populations also risks the broad genetic base of these populations, of which little is known. Lack of basic descriptive information and the probable inadequacy of terrestrial ecological theory for marine purposes, combined with the vastness of the oceans and their basic differences from land, make evaluation of human impacts difficult, even if qualitatively they are abundantly clear. Recognizing that it is economic productivity that interests most of humanity, Dr. Omori and his colleagues point out that preservation of biological diversity can have real value, if we are to gain the most from the sea.

Dr. C.K. Tseng, longtime director and now director emeritus of the Institute of Oceanology of the Chinese Academy of Sciences at Qingdao, China, traces the development of coastal mariculture along China, emphasizing the real value of recent research and development and its parallel applications for China. From his long experience, he draws conclusions regarding the overall impact of mariculture on natural oceanic biodiversity and also shows some of the details of life-cycle complexities and environmental sensitivities of oceanic life that are revealed during the intimate research associated with mariculture.

Rounding out these several viewpoints of marine biodiversity, Dr. Rita Colwell, president of the Maryland Biotechnology Institute, and her colleague Dr. Russell Hill examine the very smallest of the marine organisms, the microbial populations of bacteria and viruses. Bacteria have had a remarkable history of inventiveness and impact in the development of the earth we know today; presumed early involvement in photosynthesis, in roles in oxidation and reduction of metals (e.g., precipitation of the great banded iron ores), in fixation and reduction of nitrogen, and in reduction of sulfate have deeply engraved bacterial influence on the history of the atmosphere, ocean, and solid earth. These influences continue as bacteria intervene in a wide range of processes that cleanse, recycle, degrade, and synthesize. Modern biological technologies are offering a whole new look at marine microbial populations, recently including viruses. This new look holds enormous promise in developing economically valuable new compounds, new food production, and other new technologies; in monitoring, documenting, and evaluating human impacts; and in understanding natural processes. These new techniques are offering, in every sense of the word, a revolutionary new view, not only of microbial populations and processes, but of all life, with its diversity and value, including oceanic life.

Thus, in developing this background for evaluating the state of knowledge and effort regarding oceanic life and its diversity, a wide view has been taken, ranging from the organic wholeness of the overall system to the cellular components that circulate within it. The value of diversity—to be found in the structure of ecosystems, in the genetic potential of organisms and communities to adapt to change, in the genetic malleability of organisms to be cultured and raised for human use or needs, or in the biologically active molecular configurations that can be discovered, extracted, or cultured—is becoming increasingly apparent. Ancient arts and knowledge, classical biology and exploration, and the newest biological and oceanic technologies can and should come together in a mutually supporting web of understanding and economic and scientific advancement.

Discussion

Societal needs are placing increasing demands on knowledge and understanding of life and its diversity in the oceans. These demands range from day-to-day decisions in coastal development or fisheries management to broad calls for much wider management, conservation, and protection. The following discussion presents a general body of agreement among a group of world scientific experts in oceanic life concerning the ability of their field to respond to these demands; while acknowledging strengths, the discussion recognizes that there are serious developing weaknesses and recommends immediate and long-term remedies.

Oceanic domains were characterized very broadly for these discussions:

- The open, "blue" ocean is driven by the wind and extends to the depth of seasonal mixing.
- The coastal, "green" ocean is more dominated by its boundaries, including the shore and the bottom, with which there is commonly a systematic linkage or interchange of nutrients and energy.
- The ocean's bottom, with its benthic life, includes the highly variable shore zone, the region that directly interacts with the productivity of the "green" ocean, and the deep bottom, which only remotely or sporadically senses the productivity above or receives its organic debris as food.
- Islands of the coastal ocean are part of the coast.
- Islands and shoals of the open ocean—with emphasis on the tropical reefs, atolls, and shallow seamounts—need separate consideration.
- So do also areas with generally persistent ice cover.
- The vast, quite homogeneous, dark, stratified, internal volume of the oceans beneath the wind-stirred upper layer completes this simple categorization.

Within this vast internal volume there exists the earth's most profound reservoir of already mobilized nutrients, which, when mixed upward, fertilize the "green" ocean and, to a much lesser extent, the "blue" ocean.

Two perceptions of the oceans are worth keeping in mind as we consider their use and their life. By human dimensions, the oceans are vast and deep; there exist roughly four humans for each cubic kilometer of sea water, a part per billion, including the large proportion of "seawater" in our tissues. From a global view, however, the oceans are really very thin and widespread; the outer, thin printed layer of paper on a desk globe is a fair scale-model approximation of their thickness or typical depths. Although it may be counterintuitive, it is not necessarily contradictory to say that the oceans are enormous, remote, difficult to study and know, and have enormous resource potential, and yet feel concern for their future without good human stewardship.

Increasing fisheries, pollution, mariculture, intentional and unintentional introductions or modification of species, and other human impacts are creating unprecedented potential threats to the normal biological interactions of the oceans.

Major developing capabilities, however, may be able to greatly ameliorate the negative trends and be an important part of good stewardship. Broadly viewed, proper mariculture may offer greater good, despite its impacts, by providing food, reducing the need for traditional fishing, and increasing incentives to avoid oceanic and coastal water pollution. New biotechnologies offer major advances in food-producing efficiencies and rearability of mariculture stocks, in new biochemical products, and in dramatically new culture technologies. These new prospects depend on and can help protect marine life.

Yet, societies now give pathetically little attention to the study of oceanic life. This is a comparative assessment, relative to the emerging demands on the field, relative to comparable knowledge about life on land, and relative to other societal endeavors, considering the importance of the oceans.

These circumstances have deep historical roots, largely embedded in the land. Informal knowledge and cultural wisdom have long permitted extensive use of the wide variety of life forms from land areas and accessible shorelines. More than two centuries ago early naturalists, most notably Linnaeus, began more formal description and classification of species, assigning scientific names that were separate from the names in the local and modern languages. Yet, although fishing is an ancient and worthy practice and has more recently been the subject of extensive oceanographic research, systematic scientific study and description of oceanic life received a very late start compared to similar studies on land. Little formal or informal knowledge of the wide oceans and their life existed until not much more than a century ago, when the science of oceanography started. Military needs of World War II greatly stimulated study of the oceans and led to an era of major oceanographic expeditions following the war. New diving capabilities also afforded personal access for study, sampling, and photography. With this surge of activity, excellent advancement took place in the field from the 1950s to the 1970s, and fine researchers were attracted to the study of oceanic life during this period. But the task far exceeds one generation.

Scientific studies of oceanic life—including the most basic capabilities for describing its variability, its geographic distribution, and its natural relationships—are seriously deficient. Just as world consciousness has been driven to the need to be concerned about the basic integrity of oceanic ecosystems, the capacity of trained people to meet the increasing demands for knowledge is diminishing at an alarming rate. It is clear that the field will not be able to meet society's expectations if this trend is not reversed. Knowledge of oceanic diversity is meager. Senior researchers and experts in describing and identifying marine organisms and studying their natural relationships and evolutionary development are aging; young people are not entering the field in sufficient numbers, nor are there the professional incentives to do so. We are near the point of losing knowledge, capability, and even existing information in a field in which many years of experience go into making fine experts. These circumstances must be set right.

On a broader front, it should be pointed out that the biological sciences are making enormously valuable advances. Pushed largely by concerns for human health, stable and very substantial funding and societal support are leading toward remarkable new capabilities. Recent and wonderful advances in population genetics and genetic manipulation, in cellular and molecular biology, and in biochemistry are welcome; but for the oceans we really are still in need of the basic knowledge, description, and understanding of oceanic life and its diversity. Advanced technical fields depend on basic biological descriptions and species identifications, just as they will also help to refine them. Moreover, the unexplored, undescribed genetic storehouse residing in the diversity of life in the oceans surely will offer many surprises and valuable insights. Recognition of the organisms, understanding of their evolutionary development and evolutionary relationships, and knowledge of their natural distributions will be essential to realizing the full potential of these impressive new biotechnologies as they are applied to the oceans. In addition, these same technologies offer much to the knowledge and conservation of oceanic life. Thus, the issue is not fashionableness in the science but, rather, the overall edifice and its fundamental strength; one cornerstone is seen to be weakening beneath the great hall of the cathedral that is still being built.

The broad oceanic categories vary widely in the proportional attention they have received and in the state of knowledge surrounding them. The open ocean's planktonic life is relatively well described only with respect to several important groups. Even in this instance, however, existing theory from land fails, and no comparable body of theory has been developed for the oceans; its development almost surely awaits a detailed series of observations to understand the community's wholeness, its reliance on available resources, and the interdependencies and functions of its members. Surprisingly large diversity within the deep parts of the mixed layer is only recently being found; here, basic description still falls far short. The coastal shoreline seems best studied and understood, although its description is a moving target, as it is subjected to significant human and natural impacts. Mid-ocean reefs and shallow seamounts present a very different picture; they remain vastly underdescribed, especially with respect to their biogeography and how their life forms partake of many of the fundamental aspects of the open ocean. Curiously, some earlier studies of reefs were undertaken not so much because of their present status as elegant complexes of life but because ancient geological analogues are good oil reservoirs. Portions of oceanic reefs and atolls below common depths of diving are virtually unexplored territory. The "green" and productive coastal ocean, together with its interdependence on the estuarine zone and its bottom, is receiving growing attention, although much work needs to be done because of its local importance. There are only dozens of well-described samples of the deep bottom; even so, surprisingly large diversity is showing up, but it is still probably too early to even project overall diversity. Bottom photographs offer little help in this regard because many of the small organisms live in the thin muddy layer beneath the bottom. The deep waters are sparsely sampled and studied; broad uniformity of the deep water masses may reduce the apparent scope of this task, considering the enormous volume to be studied. As a generality, photographic/video images may contribute greatly to study of fragile floating creatures, for which gentle sampling is difficult. Finally, with regard to the very large proportion of microbes that cannot be cultivated, marine microbial diversity is effectively a new field within all of the broad oceanic categories considered.

Thus, large gaps in knowledge exist. Some major aspects of marine ecosystems do not seem to fit at all well with precepts and theories derived from land; underlying facts must be obtained and more suitable concepts developed for the oceans. The oceans are not uniform, either in space or in time; the open ocean is more uniform than inshore waters because of coastal influences. Local variability and change can give character and flexibility to local resource issues. These circumstances must be kept in mind as various concepts for stewardship and management of oceanic life are developed and evaluated.

It is important to close these gaps, both for the advancement of knowledge in the field and for the wise utilization of the oceans and their resources. Both broad and specific recommendations are set forth:

- A clear and broad need exists to review, organize, and disseminate the most current information and classifications of marine organisms, their natural relationships, and the development of their speciation and distribution.

- The study of oceanic life is ready for a *recapitulation of knowledge* and a *renaissance,* driven by needs, opportunities, and new technologies.
- Study and professional development to expand description of biodiversity and biogeography and life's natural relationships in the oceans are equally essential with these same issues on land, and they must be improved if global ecological conditions are to be scientifically predictable and managed wisely.
- Renewed attention must be paid in all areas and institutions of biological sciences and public policy to ensure that ocean life sciences are supported at sufficiently greater levels to permit existing research expertise to be applied to global-scale problems and to assure the continuance and refinement of this expertise.
- A shift in priority toward basic data gathering, description, and dissemination must be undertaken for the ocean biological sciences to develop better understanding and appreciation and to develop a sound policy for managing ocean life's abundance and diversity in light of its related social, environmental, and economic implications.
- Society in general and the national and international funding agencies and other sources specifically must redress the imbalance in research allocation priorities to give greater importance to studies of oceanic life and its many manifestations of diversity, interdependence, and wholeness.
- Species and within-species knowledge of biodiversity and other oceanographic knowledge must be integrated into ecosystem models and regional management regimes in systematic programs to ensure that valid information is available and used for problem-solving and policy-making.
- We must develop a comprehensive, computer-based global listing and information management system for descriptions of oceanic life—including its various manifestations of diversity, geography, and natural relationships—to capture, organize, and make available existing

information, to provide efficiency for work in the future, and to serve as a teaching, research, and environmental management capability. Current information management systems and capabilities should be examined, refined, and adapted to these needs.
- We must develop a plan for collections as an essential part of this specimen-based science and its data needs.
- Major national and international programs that are now being aggressively planned to address understanding and predicting earth's climate system from space, land, and sea offer enormous opportunities for the study of oceanic life. These programs can make possible symbiotic and cost-effective utilization of the data flow and the oceanic components of the technical and logistic infrastructure. Planning should embrace these opportunities. Systematic availability of synoptic oceanic data in operational time scales for biological studies of the oceans is assumed, but should be reviewed as part of overall planning.
- Team approaches will be increasingly important in these multidisciplinary and transdisciplinary areas, in which specialists in oceanic taxonomy, systematics, and natural history will occupy basic and integrating roles, nationally and internationally, as appropriate. Scientists from small or developing countries can then participate in the broad enterprise.
- These integrating programs and studies are rightly to be done in the international, cooperative, and traditionally scientific mode and will facilitate the availability of an accepted factual base that will help prevent otherwise poorly formulated, generalized environmental concerns and associated strife.

Priorities

Among these recommendations, there are two immediate calls for leadership in direct planning and action. The first is to initiate planning for the creation of a global, computer-based information system that embraces the fundamental descriptions and classifications of oceanic life and its natural relationships. The second is to recognize and implement the national and international opportunities to combine studies of oceanic life with the global-scale studies now being developed for understanding and predicting climate. Many scientific, societal, and cost-effective benefits can be realized by such a combination.

Both recommendations present excellent opportunities for national leadership and international cooperation.

Information System

The development of a computer-based global information system for oceanic life is seen as one of the most basic and important steps that should now be taken to foster the field; no such system exists now, and this circumstance seriously limits advancement. The system is visualized as starting with an emphasis on the needs of taxonomy and biogeography, but early advanced planning will permit it to evolve to full interactive capability with other data fields and programs in oceanic and atmospheric studies.

Such a system would unify and make available historical data sets, conserve existing knowledge, efficiently utilize expertise, and serve as a teaching and research capability to stimulate development of the continuing human resource base. It could serve the full range of technical development needs among nations, from sophisticated analysis to the preservation of rapidly disappearing native knowledge and lore. It could be a principal vector for interfacing the study and knowledge of oceanic life with the greater physical and chemical emphasis of other observational and analytical programs dedicated to global change. Modern communications is a powerful instrument of education and democratization, exposing people to global issues and opportunities; reliable information can help all sectors of societies make informed decisions regarding living marine resources and oceanic and coastal environments.

It is judged to be the most immediate and significant unmet need, limiting future advancement of the field. Modern information management technologies now exist to make a tremendously useful start and to make information readily available to users all over the world.

Issues of compilation, dissemination, training, data inventories, supporting collections, and broad compatibility all need to be addressed early in planning. A review of the most current knowledge and classifications is essential. Few technical disagreements are seen to exist; it is mostly a matter of evaluating and bringing earlier data into line with the most current knowledge and classifications. Workshops are recommended to accomplish the planning and development; topics include formats, standardization, and protocols for data entry; amounts of basic information and its organization; presentations; levels of search capability; basic reference frameworks in identity, space, and time; relationship to specimen collections; quality control; referencing of authority or authorship; use of images and color to supplement collections; effective methods of dissemination; increasingly facile identification schemes; flexibility to grow and be quickly updated; and interactive capability with related data sets and other geographic information systems. The basic need is to get the information compiled and fully and quickly accessible.

The system should aim to be global-oceanic but, to be realistic, should initially focus on some region: the concept of a nucleus that can grow, as the value of isolated data is perceived to be greater when part of a more comprehensive system, is seen as important. The area of the South Pacific Regional Environmental Program (SPREP) should be a serious candidate; an impressively oceanic region, the organisms of its once-remote lands are now receiving modern scientific attention. An oceanic information system would make a powerful complement to one for the land, and coordination should be encouraged and sought. Moreover, it is a region to study with some urgency because of its still-relatively-pristine quality; life uniquely restricted to the region should be documented.

Similarly, other pilot programs are possible, such as capturing the full data of a specialist's research on some group of organisms. With agreement on formatting and coordination, programs would be able to link together.

Teams of workers are seen as important to this effort. Agreement on classifications is best accomplished by those most familiar with the subject. More important, and particularly when engaging in complex data searches and working with other data fields, experts in data handling can offer expertise that will allow other researchers to address subjects and not limit their scope of inquiry to some level of personal computational capability; this may be very important in fully involving scientists from nations at all levels of technical development and from all disciplines. Such an approach has achieved a respected record of success in other global programs of recent decades, such as the Deep Sea Drilling Project at the Scripps Institution of Oceanography and the follow-on Ocean Drilling Program of the Joint Oceanographic Institutions, Inc., and seems very appropriate to new global programs in which transdisciplinary needs are best served by groups, each member of which brings a special value.

This work is seen as best progressing in the international scientific mode. Universities, laboratories, and museums are the respected origins and repositories of this kind of information. The significant purposes of this broad programmatic planning are to open the way to the future, bring the researchers of the world into interactive communication, and help prepare for much fuller integration and connection of studies of oceanic life with the major new programs getting under way to study the global earth system from land, sea, and space. Such a system would provide systematic access to the most valid and current information and cultivate international confidence. It would be an important element in a recapitulation of knowledge and would help foster a renaissance in the study of oceanic life.

Observation Systems

Studies of oceanic life must be more fully integrated into the programs being developed to study the global earth system and climate.

Today, we see many lines of effort focusing remarkable modern achievements and new possibilities toward understanding the oceans within the broader purpose of understanding and wisely using the whole earth. Recent advances in our knowledge of the changing geological shapes of the ocean basins and positions of the drifting continents, as well as our knowledge of ancient climates and their changes, have produced new insights concerning the entire outer part of the earth. These new and global perceptions are matched by the new view of earth from space. Numerical modeling with increasingly powerful computers allows us to look forward to full oceanic models with three-dimensional characterization in sufficient detail to perceive and understand many important oceanic features. These new data-hungry models are expected to be available in time to accept and use the continuous data to be expected from the observational programs that will be fully in place near the end of the decade. In the same sense, the envisioned data base and information system for oceanic life should focus toward the same time frame.

These new programs and capabilities offer remarkable synergistic, symbiotic, and cost-effective opportunities for studies of oceanic life. Such studies can engage the same flow of data and the continuously updated data fields as well as cooperatively utilize the technical and logistic infrastructure that will come into being with the new programs.

Similarly, studies of oceanic life offer much to the study of climate and its prediction. The oceans' broad circulations exert major but not well-evaluated influences on the cycling and sequestering of carbon dioxide (CO_2) by means of the growth, settling, dissolving, or burial of the countless small organisms with calcium carbonate shells or skeletal parts. Organic material and energy, originating in photosynthesis, that escapes metabolism or bacterial decay or oxidation gets buried. In both those cases involving either oxidized or unoxidized carbon, physical and chemical understanding can proceed to a certain degree, but the basic strength of the natural system and, therefore, of any conclusions concerning it is its biological strength and stability. Oceanic organisms may also be sensitive indicators of change. Deep-bottom communities—living and evolving in conditions that have been relatively stable despite profound climatic changes on land—may be sensitive indicators of subtle change originating at the surface. Other bottom communities are ravaged by muddy flows that sweep down from the continents or by volcanic ash falls; recovery from this style of disturbance seems almost a way of life. At the surface, complex boundaries of oceanic circulation affect the distribution of surface life; shifts in distributions or related extinctions may offer early and sensitive indications of change. Massive corals can contain long and detailed climate records; so-called microatolls—modest algal structures living in very shallow water—respond to and record short-term and small shifts in local sea level.

Mid-ocean reefs may serve as stepping-stones or roadside stops for colonizing and reproduction by organisms with drifting larval stages as they repeatedly go through the closed loop continuity of a reproductive odyssey that may span generations; little is known of the sensitivity of such life styles to change. Diversity of behavior of larval stages and other planktonic forms may offer clues to understanding diversity of adult forms or to understanding community diversity and its apparent stable wholeness, particularly in the nutrient-deficient central gyres. The full scale of physical oceanography—from viscosity of sea water to the great oceanic circulations—is important for understanding these small organisms as they drift through life, struggle short distances, and grab for food.

It should, therefore, be both cost-effective and scientifically important that studies of oceanic life and its diversity be programmatically integrated with the global climate and earth system studies. In its broad concept, this is certainly not a new strategy; it is the original essence of multipurpose expeditions and is to be found in the very first oceanographic voyage, that of the HMS *Challenger* in the 1870s. It is to be found in the broad conceptualizing behind encompassing titles, such as the International Geosphere Biosphere Program, and in planning efforts among scientists, including the impressive Global Ocean Ecosystem Dynamics planning. Such coordination will enhance the overall value of the major climate change and earth system studies.

At the same time, such coordination will enhance recognition and support for biological knowledge of the oceans and will help provide needed stability for this field. All societal sectors, public and private, should be aware of these needs and opportunities. Much basic, and in a sense still classical, work is needed in the oceans; the marriage of this work with some of the highest technology societies have to offer—in satellite sensing, in numerical modeling, in biotechnology, and in oceanic access—will offer new knowledge, new resources and net wealth, and new stewardship opportunities.

Studies of the ocean really are set apart from studies of the land for very much the same reasons that set apart the life of the oceans from that of the land. These reasons are generally logistic. They reflect the difference between a three-dimensional fluid volume—with a free surface and serious gradients of light intensity, pressure, temperature, and velocity of flow but generally subtle variations in most other regards—and the parallel circumstances on the solid surface of the land; both domains have the less buoyant and less viscous atmosphere immediately above. These differences are felt both by the organisms and by the researchers and their instruments. Observations from space offer a unifying aspect; satellites can yield information for land and sea. But sampling and direct study of the seas require specialized equipment and experience. Applications of new technologies, image analysis for statistical counting, deoxyribonucleic acid (DNA) analysis for whole-sample characterization, large-volume sampling, extension of ship-of-opportunity work to other technologies-of-opportunity—all offer new ways for advancement of knowledge.

It is the broad and encompassing nature of the indirect satellite sensing and associated numerical modeling programs involving oceans and atmosphere that makes them so attractive; it is the best truth from direct observational and analytical studies that makes them so necessary; it is the set of linkages between the direct and indirect domains and the flow of useful information to come from the broad combination that will make the overall effort so valuable.

The envisioned programmatic linkages involving oceanic life have purpose as well as structure. Issues of fisheries and their management, broader ecosystem stewardship, fisheries enhancement or mariculture, maritime transport, new technologies and installations, marine fouling, induced upwelling for energy or nutrients, waste management, and environmental protection all stand to benefit from the improved understanding. Long-term monitoring and time series of observations, as well as programs directed at specific sets of conditions, can benefit from associated three-dimensional knowledge of the oceanographic conditions. Large-scale features of ecosystems can be deduced, small-scale features placed in context, and critical ecosystems identified and studied. Basic instrumentation and other capabilities for study of coastal ecosystems can be most advantageously deployed. These are all good, solid reasons and justifications for proceeding with mainly physical and chemical aspects of the major global efforts; but large sectors of their benefits, particularly those associated with improved knowledge of oceanic life, will not be realized unless such applications are planned and the necessary strength in the parent fields is maintained and fostered.

The concept of an experiment is well established in science but difficult to accomplish in the oceans. Study of natural responses to known or predicted perturbations, either natural or human-made, become environmental experiments. El Niño-Southern Oscillation events, storm damage, new volcanic lava, episodes of activity at oceanic spreading centers, and turbidity flows are natural phenomena from which much can be learned about processes, for example, of recolonizing or repopulating. Human-made circumstances or events include temperature changes from discharges of cooling water, spills, changes in coastal chemical inputs, and specific actions and trials such as the introduction of new organisms, seeding of benign diatoms to compete with toxic algal blooms, or restorations of previously offended areas. The technical capability for rapidly responding to opportunities for study is important.

Past climatic conditions offer long-term natural experiments. Attempts to reconstruct changes in diversity and distribution of planktonic organisms that resulted from past broad temperature or associated sea-level excursions can be made, and they suggest general trends that may be associated with increasing or decreasing temperatures.

Technologies and efforts regarding access, sampling, and measurement must receive parallel attention. Just as SCUBA diving was a quantum change in visual and personal access, the next depth range—from 100 to 300 or more feet—will be important; mixed-gas diving or personal submersibles offer this promise. Remotely operated vehicles with telepresence and sensing/sampling capabilities afford access to a full oceanic range of depth. Ocean platforms, moored and drifting sensors and arrays, seafloor observatories, and dynamic buoys that transit repeatedly from top to bottom should all be viewed as future technological contributors to a wide range of observations in future coordinated activities. Full vertical profile ocean stations—as observatories extending from atmosphere to seafloor, from either floating platforms or appropriate oceanic islands—are certainly technically within reach; they suggest the desirability of sensors aloft to provide nested or hierarchical observations to supplement the satellite and surface observations. These and other styles of access offer major possibilities in the studies of oceanic life, as well as major contributions to other disciplines and activities. Coordinated justification in global-scale work may be complex, but in concept it is the same multipurpose expedition of 120 years ago.

In looking forward to these new prospects in access to and presence in the ocean, the purposes, styles, and techniques of sampling and measurement must be strategically addressed, and compatibility between the overall approach and the purposes must be assured. Time series of repeated samples and measurements are viewed as important; biological, physical, and chemical aspects need to be linked together through a sufficient space and time domain—for locations that are either stationary with respect to the solid earth or moving with the medium—to describe the systems and to develop the necessary understanding of the organisms' relations to their environment. The presence of a permanent platform of almost any size that is at or near the surface will aggregate organisms and cause local shifts in populations and their feeding dependencies; sampling would need to be outside such a zone of influence. Special study areas, such as critical ecosystems or selected reef areas, may need detailed surveys and fixed geographic references; detailed local bottom topography can be important to both science and safety. Physical features of sediments, such as large sediment drifts, may migrate and geographic references need to accommodate this; knowledge of disturbance and various degrees of biological recovery will help assess overall diversity of regions of the deep bottom and evaluate variability or patchiness. Protected marine areas can offer management and reproductive value for some organisms; for drifting or highly migratory organisms, it is difficult to imagine suitably large areas other than for very specific sites of reproduction, and broader concepts of stewardship seem more realistic. Knowledge commensurate with any conservation purpose is important for all situations. Strategic combinations of studies need to link the oceans with the shore, estuaries, drainages, and land; information systems can help mightily in these approaches.

Capabilities and Goals

In advocating renewed and sustained attention to the subject of oceanic life, along with immediate recommendations regarding systems for information acquisition and handling, major future and permanent capabilities to address long-term societal needs are in mind.

These enhanced systems will strengthen the necessary interactive development of knowledge in the basic fields that describe oceanic life forms and their natural relationships, ancestral development, and functional roles in larger natural systems. In turn, such knowledge will assist in designing better observational systems and better responses to ecological threats. The continuing interactive roles of modeling and theory, tested against observations and experimental opportunities, will help develop the needed body of theory that will be applicable to oceanic systems.

This scientific system should be seen as meshing with the many needs of society and helping to formulate well-considered responses within a larger social and economic system. Fisheries, recreational use, mariculture, threat analysis following disasters, natural systems' response and recovery from perturbations, waste management, transportation, mining, habitat restoration or creation, and sea defense issues are all good examples of subjects that can benefit. Mariculture that taps the profound reservoir of mobilized nutrients in the deeper waters, particularly those of the Pacific, should offer many advantages. Continuing evaluations of ecological and environmental quality, of resources and availability, and of strategies for protection and use will be served well, as will the enduring needs in education, awareness, productivity, and wise use.

In examining global-scale change and looking for its signals, variable time responses of these signals can be expected. Unambiguous signals of change would be desirable, however, and may be found first in oceanic life and its distribution, particularly in realms that are remote from direct anthropogenic effects. There is need to establish a basic and broad qualitative understanding of the organic response to change and to then refine it progressively to give more quantitative and more precise predictions, along with their implications and consequences, at both local and global scales. In addition to coastal zone management decisions to cope with future changes, there is surely the need for planning to recognize natural systems that overlie national boundaries.

There is, therefore, a whole range of issues for which renewed and sustained attention to oceanic life will be important.

For oceanic realms, the importance of these issues and the uncertainties surrounding them seem underestimated; the public perceptions and expectations of the level of knowledge of oceanic life that can be brought to bear on these issues seem overestimated. Simply inventing the answers will not suffice.

A communications revolution now embraces the earth; profound political changes are taking place. Communications and weather satellites are now permanent and desired fixtures in today's societies. Similarly, as improved short-term climate forecasting—reasonably to be expected from global observing systems—is developed and demonstrated, these related systems can be expected to become permanent. The oceans and their life must be part of this future view, and their study must be adequately supported and fostered as this view comes into focus. As ideological and military issues continue, as we hope, to become less important, enlarged or revised agendas of security for and among nations will increasingly confer greater importance to ecological and environmental security.

As plans are developed to systematically view the whole earth, it is important to maintain the strength and vitality of those sciences that will bring forth so much of the real meaning and value of the observations.

Whole ocean stewardship must be an ultimate objective. This implies wise use of, and a sense of good will toward and dependence on, the natural system, as well as knowledge of the system, systematic access to this knowledge, and the continuing human resource of people who know how to use this knowledge.

2
Managing Biodiversity in the Oceans
Martin V. Angel

Climate change radically affects the richness of the life around us, and the implication of such changes for human survival and quality of life is a dominant theme in global ecology. This green revolution in our thinking is still to be translated into practice, however. How is biological knowledge and theory to be put to use in managing the resources of the earth? Although the maintenance of biodiversity has become one of the central tenets of the global conservation movement, its ecological importance remains poorly supported by scientific observation and theory. With the human population predicted to grow and stabilize at around 10 billion by the year 2100, the threat to all global systems, including the oceans, is likely rapidly to reach a critical level unless adequate integrated management for all global resources can soon be put in place. The development by the United Nations Environmental Programme (UNEP) of a Global Convention on Biodiversity and the follow-on requirements and decisions for implementation and ratification have given continued urgency to the need to define what is meant by biodiversity. Yet discussions under the United Nations Commission on Environment and Development largely ignored the oceans and their importance to life on earth. In this paper some of the key issues will be discussed.

What is Biodiversity?

The concept of biodiversity encapsulates the complexity of the mosaics of life forms on earth at a number of hierarchical levels of organization. At the lowest organizational level of the individual organism is *genetic diversity*; this describes the variety of genetic information, mostly in the coding provided by sequences of bases in deoxyribonucleic acid (DNA), which flows through populations of the same species.

At the next level is *species diversity*. This has two components: first, *species richness,* which describes the number of species inhabiting any point in space and time; and second, *evenness,* which takes account of the relative abundances of the species within the community (or assemblage) and the degree to which a few of the species may be numerically dominant.

Ecosystem diversity describes the patterns of the mosaics formed by the various types of assemblage. Detecting and quantifying the mosaics depends on the use of appropriate sampling scales, both temporal and spatial; otherwise, the determination of whether or not assemblages are similar either in composition or ecosystem function may be flawed. The term community is often loosely applied to the species composing the assemblage observed at a given point in time and space. However, all the observed species will not necessarily be "regular" in their occurrence or play an active role in the ecosystem's dynamics, nor will all the "active" species necessarily be collected. Ideally, "community" should be used to include only those species that have a regular functional role in the system's dynamics.

Across these three organizational levels of diversity lies another, *physiological diversity.* This describes the rich variety of functional architectures and physiological adaptations assembled through natural selection modifying the limited range of cellular building blocks and solving the problems of survival. Many of these solutions have anticipated human ingenuity and technology by millions of years.

If oceanic biodiversity can be considered to be a resource, important to be maintained by environmental management, then a range of questions arise. At the species level there is the need to determine exactly what factors control diversity. The dynamics of ecosystems have to be understood—what controls migration, competition, food-web structure, speciation, and extinction?—all of which are factors that strongly influence ecological processes and have practical implications for resource management. For example, is diversity of marine ecosystems determined top-down (through the influence of top predators) or bottom-up (through the impact of environmental factors determining the composition of the phytoplankton)? How in the past has the removal of top predators (whales or commercial fishes) affected the structure and the dynamics of the communities? Which (if any) anthropogenic chemical inputs—through direct discharges into the ocean or indirectly into rivers or via the atmosphere—have resulted in sufficient shifts in chemical cycles to induce changes in ocean communities? Do we know to what degree contemporary human activities are inducing changes in the oceanic environment and diversity, and how can we improve predictions of the impacts of future activities?

At the ecosystem level there are equally difficult problems to be resolved. Why are some areas so uniform while others are variable? What determines the coarseness of the "graininess" of the mosaics formed by and within the ecosystems? What determines the sharp shifts in community structure seen along latitudinal sections in the oceans, and how are these reflected in process rates and material fluxes? How do variations in productivity at all scales of time and space influence community structure?

Why Are the Oceans Special?

The oceans cover 70 percent of the earth's surface to an average depth of 3,800 meters (figure 1). Primary productivity, restricted to the surface few tens of meters, feeds almost all life in the oceans. On average, the oceans are not as productive as the land, but they are so vast in area that photosynthesis by oceanic phytoplankton accounts for nearly half the global total of carbon fixed annually. A substantial portion of this fixation is by picoplankton and nanoplankton less than 5 micrometers in diameter, so their annual fixation of carbon dioxide exceeds that fixed by tropical rain forests.

The oceans contain relatively few species. Sournia and Ricard (1991) estimate that there are only 3,500 to 4,500 species of oceanic phytoplankton (in contrast to an estimated more than 250,000 species of terrestrial plants). Moreover, the sedimentation and subsequent sequestration in oceanic sediments of carbon fixed by these few species in the surface waters of the ocean into deep water must serve to slow down

Figure 1
Hypsographic curve showing the relative proportions of altitude of land surfaces and bathymetric depths in the oceans

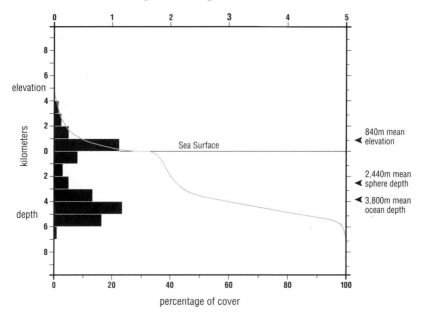

the rate at which carbon dioxide concentrations are increasing in the atmosphere. In terms of global processes, which is the more important—a species of oceanic picoplankton or a rain forest tree? Can we establish whether there is any anthropogenic threat to these oceanic species?

Despite their relative poorness in numbers of species (true for pelagic species but too little is known about benthic species to be certain), the oceans are exceedingly rich in higher taxa. They harbor 28 phyla of animals, 13 of which are endemic to the oceans, compared with the 11 phyla that inhabit terrestrial habitats (only one endemic; see table 1). In paleontology this richness in higher taxa is known as disparity (Jaanusson 1981, Runnegar 1987). The maintenance of taxonomic disparity in this sense has been overlooked by most conservationists. In establishing methods for objectively assessing priorities for

Table 1
Distribution of animal phyla by habitat
Each phylum restricted to a single habitat (i.e., endemic) is indicated by ±.

	Marine	Symbotic	Freshwater	Terrestrial
Acanthocephala		±		
Annelida	+	+	+	+
Arthropoda	+	+	+	+
Brachiopoda	±			
Bryozoa	+		+	
Chaetognatha	±			
Chordata	+	+	+	+
Cnidaria	+	+	+	
Ctenophora	±			
Dicyemida		±		
Echinodermata	±			
Echiura	±			
Gastrotricha	+		+	
Gnathostomulida	±			
Hemichordata	±			
Kamptozoa	+	+	+	
Kinorhycha	±			
Loricifera	±			
Mollusca	+	+	+	+
Nematoda	+	+	+	+
Nematomorpha		±		
Nemertea	+	+	+	+
Onychophora				±
Orthonectida		±		
Phoronida	±			
Placozoa	±			
Platyhelminthes	+	+	+	+
Pogonophora	±			
Porifera	+	+	+	
Priapula	±			
Rotifera	+	+	+	+
Sipuncula	+			+
Tardigrada	+		+	+
Totals	28	15	14	11
Endemic	13	4	0	1

conservation, however, Vane-Wright et al. (1991) have used cladistics to weight species priorities according to their evolutionary disparity (see below). This high degree of disparity in oceanic faunas probably arose from life originally evolving in the proto-ocean; relatively few phyla succeeded in making the transition between aquatic and terrestrial habitats. The deep ocean remains a refuge for many archaic albeit specialized forms; these endemic and archaic taxa will no doubt hold many genetic surprises.

It is also worth bearing in mind that compared with rarity in terrestrial habitats, a rare species in the oceans may be globally very numerous because of the sheer size of the habitat. Similarly, because the abundant pelagic species occupy a three-dimensional habitat, their global abundances will be far larger.

Thus, in terms of global processes the oceans are just as important as terrestrial systems, although of less immediate and direct importance to mankind and, at present, less threatened. Oceanic metazoans are less diverse, more disparate, and far more numerous than their terrestrial counterparts. Many appear to have persisted much longer through geological time.

Global Variability in Time and Space

On earth, species are distributed in chaotic mosaics both in space and in time. At any specific point the assemblage of individuals includes both long-term and temporary inhabitants, only some of which are contributing to the local processes and can be considered to be components of the community. At large spatial scales the composition of assemblages shows patterns of latitudinal regularity and continuity. Some researchers believe that by using various techniques of analysis (e.g., cladistics), they can still detect echoes of past geological events such as the fragmentation of the supercontinent Pangaea, the opening and closing of the Panamanian isthmus, and the repeated drying out and inundation of the Mediterranean. There are also memories of more recent climatic events driven by planetary cycles with periodicities of 10^5, 35×10^3, and 23×10^3 years, which lead to major shifts in climate such as the cycles of glaciation. The forcing created by these long-term cycles and events is clearly seen in the geological sedimentary record read by the Ocean Drilling Program (ODP) and its predecessor. The sediment sequences, however, contain a readable record from a very limited selection of taxa, mainly foraminiferans, diatoms, radiolarians, coccolithophorids, and silicoflagellates. Apart from the pelagic foraminiferans, these are all groups whose present-day biogeography and ecology are relatively poorly known. The record appears to be fully consistent with the biogeographical patterns of species distributions following the same pattern as the large-scale circulation patterns of global currents, as shown by the water mass distribution and also reflected in many chemical properties of the ocean waters.

The large-scale patterns of ocean circulation are determined by planetary forcing—the rotational effects of the earth on the atmosphere and the transfer of energy from the wind field to the sea surface. These patterns are modified by the shapes of the ocean basins. So, as the configuration of the continental land masses in the geological past has become known, the past biogeographical patterns of the pelagic communities can be inferred and confirmed by the fossil record. Equally revealing are the estimations of past surface and bottom water temperatures derived from the stable isotopic ratios of carbon and oxygen in skeletal carbonates of surface-dwelling and benthic-dwelling taxa. The temperature of the deep water of the open oceans until the Middle Eocene about 50 million years ago was quite warm, about 8–10°C and only slightly cooler than the present deep water in the Mediterranean. In neither of the seas with warm deep waters has a bathypelagic fauna evolved: although the Mediterranean has been subject to repeated cycles of drying and re-flooding, the Red Sea has a long enough geological record for such a fauna to have evolved. Have cool bottom waters been a prerequisite for the evolution of specific bathypelagic and abyssopelagic faunas? The depth at which calcium carbonate dissolves—the lysocline—was much shallower in the geological past, which will have limited the use of skeletal carbonates to much shallower depths than in the present ocean. The separation of the Australasian land mass from Antarctica created the conditions for the formation of the circumpolar circulation in the Southern Ocean and for the formation of cold bottom water. Another curious factor is that the strange anchialine faunas of saltwater caves in Bermuda and the Bahamas that may have been isolated since the formation of the islands some 60 million years ago have their closest relatives in the abyssal ocean. Is that a coincidence or an indication that the invasion of the deep ocean by pelagic faunas only became possible as the deep water cooled?

At shorter time scales, catastrophic events may have resulted in local (or even global?) extinctions that may then have been followed by immigrations and emigrations. Diatom extinctions 8 to 10 million years ago and about 2.5 million years ago were associated with cooling events (Barron and Baldauf 1989). The historical records of the Northern European sea tell a similar story of long-term variations that have had major impact on local communities—the catastrophic decline of the Hanseatic League, which was based on a herring fishery that collapsed at the beginning of the Mini-Ice Age, and the mass starvation in the Faeroes in the early seventeenth century when the cod fishery totally failed (Lamb 1977). There are sharp and sudden changes in the abundances of key species, for example, the occasional population outbursts in abundance of the crown-of-thorns starfish, the switches seen in the Russell cycle in the North Sea, and the oscillations seen in dominance of commercial fishes in the California Current. These may be triggered by decadal scale events such as the "Great Salinity Anomaly" (Dickson et al. 1988) that was first observed in 1969 off Greenland and then was followed along a tortuous trajectory around the North Atlantic until 1981/82 (figure 2). The varved sediments found in some of the anoxic Californian basins have revealed long-term cycles of change that are still to be correlated with environmental variations. El Niño-Southern Oscillation events (El Niños) create approximately decadal perturbations that have the potential to influence the community structure of organisms with short life cycles (of a few days or weeks) but may have less effect on the longer-lived species. The data produced for the oligotrophic regions of the North Pacific by McGowan (1990) show that El Niños have had no effect on the species composition or domi-

Figure 2
Track and timing of the circulation of the "Great Salinity Anomaly" around the northeastern Atlantic from 1968 to 1982

The figures along the track indicate the estimated deficiency in 10^9 tons.

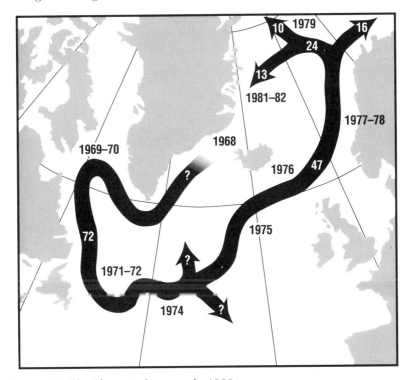

Source: Modified from Dickson et al., 1988.

nance during the last decade. As well as interannual variations there are seasonal—intra-annual—changes in community structure that reflect the different ways whereby individual species cope with seasonality—by migration, hibernation, aestivation, or diapause.

These variations in time and space are often conceptualized in the form of Stommel diagrams (figure 3). Because hydrodynamic features tend to have an approximately linear relationship between their size and their persistence in time, these diagrams give a useful framework for discussions of the causality of the observed patterns.

Figure 3
Stommel diagram for zooplankton biomass variations in time and space

The diagram illustrates the strong correlation between the biomass and the hydrodynamics of the oceans.

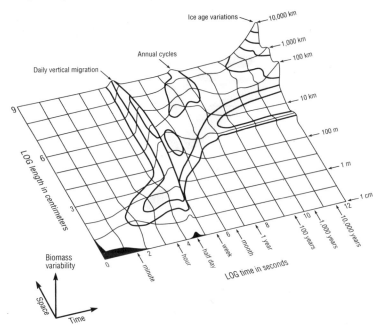

Source: Adapted from Haury, McGowan, and Wiebe 1978.

The Species Concept

The species concept is fundamental to how we classify life on earth and to our concepts of the dynamics of community structure and function. It is far from perfect, however, particularly when applied to many microbial and lower plant groups. Some assemblages and their component species have been highly persistent and stable; others have changed continually. Conservationists tend to focus attention on stable communities, which are often richer in species and include more rare species. They seldom address how ecological criteria should be used to determine priorities for conservation and management based on processes, especially processes on a global scale.

Species are not static entities. There is a hierarchy of response to environmental change. There is toleration by acclimation, avoidance by migration or entering a resting phase, and genetical adaptation leading eventually to speciation. If the degree of change exceeds the limits of individual tolerance (this will include a time factor), then death will ensue. Genetic heterogeneity within a population will result in the lethal limits varying within a population, but the ultimate result of too large and/or too rapid change is local and ultimately global extinction.

The degree of tolerance shown by species is correlated with the environments they inhabit. In seasonal environments, which tend to be more variable, the inhabitants tend to be physiologically flexible and more tolerant of fluctuating conditions. In environments that are variable over geological time scales, such as temperate latitudes subject to the glacial cycles, species tend to be genetically more flexible and, in response to even quite rapid changes, readily diversifying to become new forms (i.e., speciate). The classic example is provided by the freshwater fishes of the African lakes that, as a result of fluctuating water levels, are thought to speciate within a few hundred years. In environments that have remained constant or highly predictable over long geological periods, species tend to have become much more specialized to the limited range of conditions and to be intolerant of change. Such species may lack needed resilience and so face extinction if they are subject even to quite gradual change. Such change may not necessarily be the result of changes in the chemico-physical environment; the biotic environment may change either through an existing component species gaining or losing its competitive edge (even becoming extinct) or through the introduction (natural or anthropogenic) of new competitors. Such changes may have far-reaching impact on the patterns of competition, food-web structure, and energy flow through the ecosystem.

In the pelagic oceanic ecosystem there has probably been considerable continuity over geological time, especially in deep water. There have been considerable changes in the geographical locations of oceanic fronts and boundaries, but the communities have probably moved with the conditions. Thus mesopelagic and bathypelagic communities can be expected to contain the most highly tuned species with possibly the least tolerance of environmental change of all on earth.

How the Diversity Concept Developed

The concept of biodiversity is most highly developed at the species level. Ever since Linnaeus's protocols for the naming and cataloging of species (taxonomy) were universally adopted, the task of developing a global inventory has continued. It soon became evident that species are to be found in greater profusion in some places than in others, giving rise to the science of biogeography. This descriptive exploration has progressed relatively slowly in the oceans because of the difficulties of sampling in remote and hostile environments and the oddity of some of the inhabitants. New habitats with totally novel faunas are still being discovered as our sampling skills improve (e.g., hydrothermal vents and cold saline seeps).

A major impediment to progress in the systematics of some taxa has been the need for special techniques. The taxonomy of microorganisms made little progress until the invention of electron microscopy; the full impact of the more recent development of molecular biological techniques is yet to be realized. Until recently no one believed that there are viruses in the marine environment, and their role in structuring communities and regulating patterns of succession is a rich vein for speculation with little hard fact. The richness of the variety of fragile gelatinous organisms was only fully recognized once manned submersibles became freely available and techniques for their gentle collection had been devised. Although the problems of their taxonomy are slowly being sorted out, their role in pelagic communities remains unresolved.

Taxonomy is more than just an "artificial" cataloging system for classifying the wealth of living forms; it should be "natural" and reflect the evolutionary relationships between species. Natural classifications are difficult to develop, however; often this is because distantly related taxa have converged in form through their adoption of similar solutions to the problems of survival in a particular habitat. Good natural classifications or systematics are a prerequisite for correctly interpreting some of the results and applications of "high-tech" new sciences such as molecular biology and genetic engineering. A major stumbling block to progress is the relatively disorganized state of existing taxonomic knowledge. There have been few attempts to develop comprehensive inventories of described species. Even in those taxa for which species listings and/or data bases have been developed, the lack of compatibility between information systems inhibits the accurate compilation of existing knowledge.

Ecosystem diversity had its origins in botany at the turn of the century. Many of the basic concepts of community, classification and ordination, and scale and pattern had their origins in quantitative plant ecology (e.g., Greig-Smith 1964). Terrestrial zoologists developed the concept of food webs, and Elton (1966) postulated the theory of ecological pyramids whereby secondary production is partitioned between herbivores, primary carnivores, and secondary carnivores (in descending order of turnover); this concept continues to pervade terrestrial ecology. Pelagic oceanic ecosystems, however, tend to contain "unstructured" food webs (see Platt et al. 1981) whose structure and dynamics are determined more by size than by species composition; little "particles" are consumed by large "particles" almost regardless of their "species." Thus, a newly hatched herring larva initially runs the danger of being eaten by carnivorous plankton, but when it has matured, it will eat its former predators. A terrestrial equivalent would be a mouse that grew eventually to a size at which it would start to eat cats—a sort of "Tom and Jerry" world! Thus the ecological theories and concepts that have been developed for terrestrial systems do not necessarily translate well when applied to oceanic environments.

There are similarities between oceanic and terrestrial communities. More species tend to occur at low latitude than at high (figure 4). However, whereas species richness decreases with altitude on land, it initially increases with depth in the oceans (to depths of 500 to 1,000 meters)—see below—even though the communities are becoming increasingly remote from their primary sources of energy in the sunlit

Figure 4
Variations in numbers of species collected in the water column to depths of 2,000 meters

The species were collected at a set of standard stations in the Northeast Atlantic approximately along 20°W.

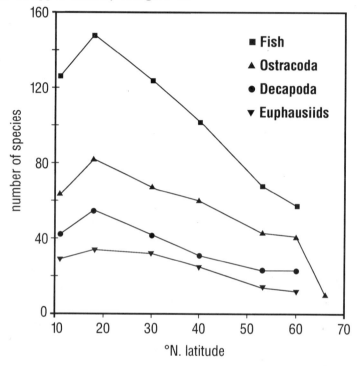

Source: Based on data in the Institute of Oceanographic Sciences (IOS) pelagic data base.

euphotic zone. Both on land and in the sea, average animal size tends to be greater at high rather than low latitude. Average size tends to increase with depth for invertebrates, but the relationship is complicated by the changes that occur when the dwindling food resources result in the disappearance of certain taxa and their replacement by others.

An important concept in terrestrial ecology arose from the empirical observation that, although the inventory of species inhabiting an "island" may fluctuate as a result of immigration and local extinction, their average number tends to be correlated to the island's area. This theory of "island biogeography" (MacArthur and Wilson 1967) became extended to cover all isolated habitats, from mountain tops to remnant pieces of forest. In pelagic environments in the oceans, the dominant factor in determining the distribution of communities is the large-scale hydrodynamics (e.g., McGowan 1990). The Stommel diagram (figure 3) shows that there are no "islands" in the terrestrial sense in the open ocean. Smaller-scale features, while generating patchiness, do not appear to be important in determining overall community structure because of their short time scales. There are deep basins and isolated seas that contain isolated subpopulations and so evolutionarily may function as islands. For benthic communities, bottom topography is important; also, interactions between organic inputs, which are determined by the overlying hydrographic regime, and the benthic boundary layers serve to create a more heterogeneous community structure than exists within the water column. The communities associated with hydrothermal vents, hydrocarbon, and cold saline seeps are probably special cases. Their distributions are disjunct and the features around which the communities cluster are ephemeral (Tunnicliffe 1991). The local productivity is greatly enhanced through chemosynthesis, and this is expressed in the standing crop being orders of magnitude higher than in the surrounding regions. The communities tend to be species-poor and to be dominated by a very few taxa. The zoogeography of these unusual species is poorly understood, partly because these communities have only very recently been discovered, but also because finding and investigating these small and remote oases in the otherwise barren expanses of the abyssal ocean require heavy investment of time and expensive technologies.

Close inshore interactions between the topography and geology of the seabed, tidal currents, and riverine outflows make for a finer-scale heterogeneity of the physical environment, which is reflected in the heterogeneity of the biological communities (e.g., Ray 1991). This has considerable implications for the problems of management in coastal versus oceanic waters; relatively small-scale management as practiced in terrestrial environments by establishing reserves and management zones can be effective inshore (e.g., Ballantine 1991) but not offshore.

While there is a tendency to adopt terrestrial ecological theory uncritically in the study of oceanic ecosystems, there is also a slowness in discarding invalidated theory. For example, there was an intuitive hypothesis that food webs with more cross-links are more stable than simple systems. That hypothesis has been shown to be fallacious, and simple ecosystems have been demonstrated to be more robust. Yet the argument still prevails that the Southern Ocean ecosystem is fragile because it is simple, with krill occupying a "keystone" position in the food web. Indeed, the system appears to be robust, having apparently withstood the removal of its top predators—fur seals and large whales. However, although there are extant collections whereby this apparent robustness can be examined, the critical study has not been undertaken.

Biodiversity in the Oceans

So where does our knowledge of oceanic ecosystems fit within the concept of biodiversity? Empirically, diversity tends to increase as the chemico-physical environment becomes more stable and/or predictable. Diversity tends to be correlated with productivity and latitude. For example, Southern Ocean assemblages are less species-rich than equatorial assemblages, but they are richer than their Arctic counterparts.

Within the relatively few taxonomic groups for which there is good global coverage, there are correlations between large-scale patterns in pelagic distributions and the gross features of global circulation (van der Spoel and Pierrot-Bults 1979; McGowan 1990). The ranges of individual species, however, rarely coincide precisely with the main hydrographic boundaries. This may result from differences in the survival of the expatriate individuals that advected within eddies and filaments into regions where their populations cannot continue to persist, or it may result because there is no direct correlation between the hydrographic features and the limits of survival and persistence of the individual species. More precise data are particularly required for those groups that become incorporated into the fossil record—foraminiferans, coccolithophorids, radiolarians, pteropods, and diatoms, whose sediment records are used to make extrapolations of past climates. These are taxa of quite low diversity and wide environmental tolerances, and their distributional boundaries are blurred by oscillations in the major circulation fronts and the effects of mesoscale eddies.

Haq (1984) illustrated how ocean circulation, upwelling, and bottom water formation have shifted as continental drift resulted in the breakup of the supercontinent of Pangaea into the present configuration of continents (figure 5). Van der Spoel et al. (1990), in using a cladistic approach to the systematics of euphausiids and hydromedusae, detected memories of these ancient oceanic gyres in the present distributions of the more archaic taxa. Perhaps less controversial are the analyses of those differences between the Pacific and Caribbean faunas that have developed since the Miocene closure of the Panamanian isthmus. There is a large pantropical fauna that has persisted, presumably with little change, despite the disruption in gene flow for the last few million years. This implies that speciation in the oceanic species is relatively slow, certainly compared with the cichlid faunas of some of the African lakes in which speciation is thought to have occurred within 500 years of isolation.

The mechanisms of speciation remain controversial. Isolation is considered by some to be a prerequisite, i.e., disjunct populations with no gene flow can result in allopatric speciation. Others argue that speciation may also occur sympatrically—i.e., within populations that geographically overlap. There are taxa that speciate more readily than others, and this is correlated to some extent with life-history characteristics. In the present ocean, it may seem difficult initially to see how speciation can possibly be allopatric. The global ocean turns over every 1,500 years, and the Atlantic every 250 years; at such stirring rates, in the absence of limitations, all species would be ubiquitous within a millennium. However, this ignores the influence of the glacial cycles.

Paleoclimatic studies (Cline and Hayes 1976) showed how pelagic distributions have shifted between the height of the last glaciation 18,000 B.P. and the present. The distribution of sea-

Figure 5
Six stages in the evolution of the ocean basins during the geological past showing the probable circulation pattern created by planetary forcing

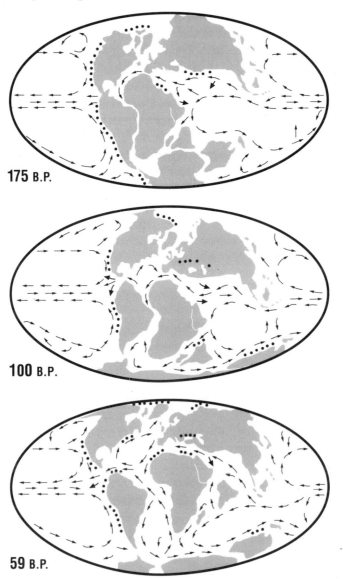

175 B.P.

100 B.P.

59 B.P.

The numbers to the left of each diagram indicate the time in millions of years B.P. The arrows indicate the likely regions of bottom water formation, and the dots indicate regions of coastal upwelling.

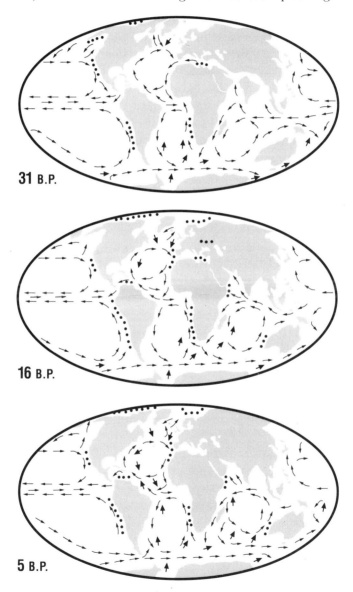

Source: Modified from Haq 1984.

surface temperatures illustrates the dramatic changes that took place (figure 6). In the Mediterranean (and in the upwelling region of Northwest Africa) relict populations exist of cool-water, higher-latitude species such as the lantern fish *Benthosema glaciale* and the euphausiid *Meganyctiphanes norvegica.* These populations not only remain isolated from their parent stock, but in the Mediterranean they are now living in warmer waters (higher than 12.7°C) than the parent stock will encounter for the most of the year. In the North Atlantic, the adults are diel vertical migrants that move up to the seasonal thermocline but not into the wind-mixed layer during the summer. South of 40°N the North Atlantic stock

Figure 6
Comparisons of summertime sea-surface temperatures in the Atlantic between the present (A) and 18,000 B.P. (B) at the height of the last glaciation

Note how the changes were more extensive in the Northern Hemisphere with the Polar Front having migrated to its present position north of Iceland from 45°N.

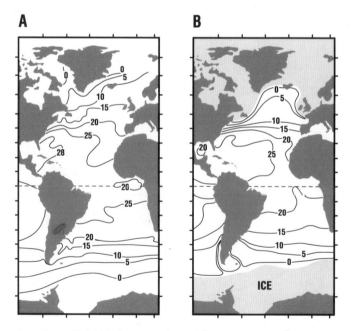

Source: Based on CLIMAP data as adapted from various references.

submerges to inhabit depths of 1,000 meters or so, becoming bathypelagic. When, and if, these presently isolated populations are reunited during the next glaciation, they may well have diverged sufficiently to have become different species. If this mechanism of isolation is valid for oceanic species, then those regions in which there are deep basins with shallow sills can be expected to be centers of speciation as a result of fluctuating sea levels during the glaciations. The region of the East Indian archipelagos should be expected to be particularly diverse, and Fleminger (1986) attributed the high richness of some copepod genera to the fluctuating changes in sea levels (figure 7).

Figure 7
Hypothetical sequence of speciation in the copepod group *Labidocera pectinata*

The diagrams indicate the hypothetical sequence of speciation during the succession of sea-level changes since the Pliocene that created and removed year-round and seasonal barriers to the distributions of the sibling species. The different patterns indicate different species.

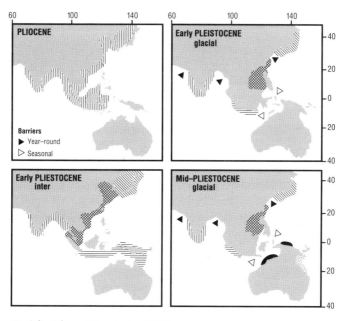

Source: Modified from Fleminger 1986.

44 Diversity of Oceanic Life: An Evaluative Review

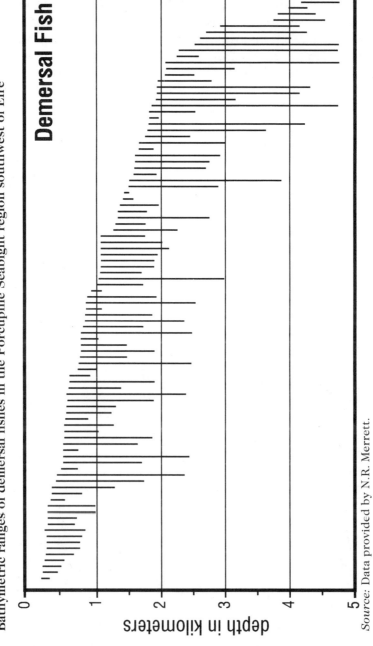

Figure 8
Bathymetric ranges of demersal fishes in the Porcupine Seabight region southwest of Eire

Source: Data provided by N.R. Merrett.

There are consistent trends for species richness to increase in both benthic and pelagic animal species with increasing depth down to 500–1,000 meters. Figures 8 and 9 show examples of the bathymetric ranges of demersal fish and planktonic ostracods. Figure 10 shows depth variations in species richness for planktonic ostracods at various latitudes in the Northeastern Atlantic. Figure 11 shows the observed data for four megabenthic taxa of southwest Ireland in which maximum diversity tends to be at around 1,000–2,000 meters. This enhancement of species richness with depth runs counter to the exponential decrease in community standing crop; standing crop at 1,000 meters is about an order of magnitude lower than at depths of 0–100 meters (Angel and Baker 1982). In the top 1,000 meters, pelagic communities have a zonation, related to the light field, that has wide geographical continuity. Many of the morphological adaptations shown by the inhabitants of the mesopelagic and bathypelagic zones appear to be bizarre but on investigation prove to be elaborate devices for survival, feeding, and reproduction. Many of these adaptations are depth specific, for example, the high degree of silvering seen in certain myctophid and gonostomatid fishes limits the species to daytime depth ranges of around 300–700 meters (the precise range is dependent on the transparency of the water). There are many examples in the uses of bioluminescence. There are reports of similar zones in benthic and benthopelagic species, but these now seem to be the product of sparse sampling. In demersal fish from depths greater than 200 meters to abyssal depths in the North Atlantic there is no geographical continuity between the putative zones (Haedrich and Merrett 1990).

Phytoplankton also present problems of interpretation. According to ecological theory, coexisting species should partition resources, each having its own "niche" with minimum overlap in their environmental requirements. Once an ecosystem has stabilized and reached equilibrium, interspecific competition should be minimized; jostling by natural selection occurs mainly when outside species invade or barriers to isolation are removed. In the tropical Pacific, Venrick (1986) found more than 200 phytoplankton species inhabiting

Figure 9
Bathymetric ranges of planktonic ostracods

The diagram indicates the bathymetric ranges of planktonic ostracods in the water column at 42°N 17°W, together with an indication of the depth ranges of the classical zonation of pelagic communities: (1) epipelagic; (2) mesopelagic; (3) bathypelagic; and (4) abyssopelagic. Dashed lines indicate decreased certainty.

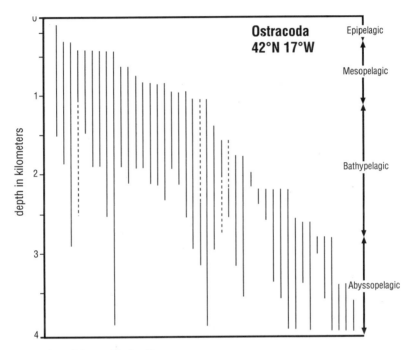

Source: Angel (unpublished).

the turbulently mixed euphotic zone, stretching credulity in the niche hypothesis. She found two main phytoplankton assemblages, one in the high light/low nutrient regime of the wind-mixed layer, the other in the low light/higher nutrient regime in the nutricline. No one species dominated, and consistently the majority of species were rare. In the same locality, McGowan and Walker (1985) found 175 species of copepod with just seven patterns of vertical distribution. Once again, the majority (more than 100) were consistently rare in sampling, repeated in different seasons and years.

Abyssal meiofaunal populations are very heterogeneous at fine scales (1–10 centimeters), but as the scale of observation increases the pattern becomes more uniform. Similar mosaics occur in tropical rain forests created by patterns of tree fall (Whitmore 1982). The graininess of the mosaics in benthic communities may, as in tropical rain forests, be determined by natural perturbations, which may range in scale from disturbance caused by a burrowing infaunal animal to the extensive devastation caused by turbidity flows and slumps. The communities may never reach a state of equilibrium, but remain a dynamic mosaic of assemblages at different stages of recovery. Individual species have different response rates to perturbation and new opportunities, so chance plays a significant role in what lives where. Experimental studies have demonstrated that there are pioneer weedlike species that are eventually suppressed as the longer-lived, slower-growing species reestablish. Conversely, long-lived species may be able to persist for decades when conditions prevent them from breeding successfully, misleadingly giving an impression of stability. It is still not known how the inhabitants of ephemeral habitats, "wood islands" (Turner 1981), and hydrothermal vents (Tunnicliffe 1991) locate new habitats so quickly.

McGowan (1990) summarized two decades of his paradigm-making studies in the Pacific. He concluded that, although populations of pelagic species are often highly dispersed in terms of numbers per unit volume, their vast ranges result in their populations being very large. The high species richness of pelagic communities results from the occurrence of many rare species. In Pacific oligotrophic central gyres, the dominance structure of communities has remained consistent over decades, despite the marked reduction in the numbers of top predators (tunas); these ecosystems have long-term stability. The hypothesis of resource allocation as the mechanism controlling the structuring of the communities is seldom upheld. Where there is a large advective signal (e.g., the California Current) there are large local changes in dominance structure. Despite the significant departures from long-term climatic means of environmental variables during the past two decades (resulting from El Niño events and in-

Figure 10
Profiles of species richness of planktonic ostracods

The diagram shows profiles of species richness of planktonic ostracods in daytime samples collected approximately along 20°W in the northeastern Atlantic, showing that maximum richness is consistently between 500 and 1,000 meters and that the reduction in species richness at latitudes more than 40° is expressed throughout the water column.

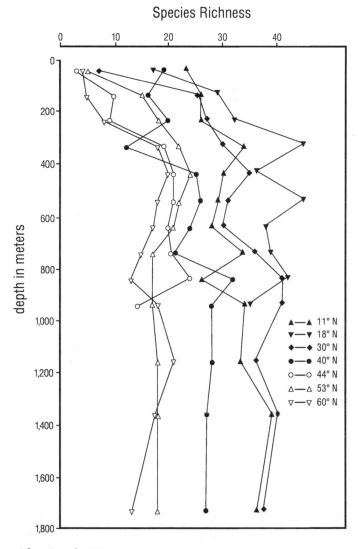

Source: After Angel 1991.

creased pollutant inputs), no pelagic species are known to have gone extinct, and there have been no persistent changes in the structure of the ecosystem. The problem of understanding the role of the rare species remains. They are a constant feature of pelagic systems, and it must be assumed that they play some role. Yet present theories of diversity cannot account for their presence, so the theory is either flawed or inappropriate for application to pelagic systems. This is the same body of theory on which policy is now being developed for ocean management and conservation!

Biodiversity and Environmental Management

What are the key concepts that are emerging? Management and conservation have to be the practice of the possible. Extinction is a fact of life that will continue no matter how skillful we become in the art of preservation. The *cri de coeur* once heard at a major international conservation meeting that "allowing any species to go extinct is an immoral act" failed to recognize that extinction is a fact of life. No species is immortal, although it is argued that the information in the DNA molecule is as close to being immortal as any "living" system. One consequence of this belief in the importance of individual "flagship" species is an unbending emphasis by conservationists on rare animals and their disregard of the commoner species that play a central role in ecological processes. Rarity is equated with vulnerability. McGowan (1990) has cogently argued that rarity in pelagic species is the norm; so many pelagic species (maybe 80 to 90 percent) are consistently rare that it would be most unwise to assume they play a negligible role in ecosystem functions; yet terrestrial theory leaves no room for rare species to have any significant ecological role. The ecology of rarity needs to be understood better. Otherwise, investment of the limited resources for conservation may be expended on the protection of individual species that are either at high risk of going extinct (locally or globally) at any time or, at the other extreme, are far from being inherently vulnerable. Similarly, the protection of unstable ecosystems would seem to be an unwise use of limited resources. Small isolated populations will always be susceptible to local

Figure 11
Species richness of megabenthos taxa at depth intervals of 500 meters down the slope of the Porcupine Seabight

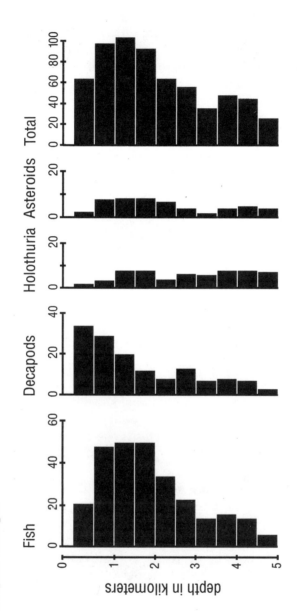

Sources: Data for fish derived from figure 4. Data for decapods, except for pagurids, provided by Dr. A.L. Rice (note that the data for brachyurans—the crabs—contributes to the shallowness of the peak of species richness). Data for holothurians (sea cucumbers) from Dr. D.S.M. Billett. Data for asteroids (starfish) also from Dr. D.S.M. Billett.

catastrophic events—pandas are at risk when all the bamboo flowers at once, tuataras are vulnerable to the introduction of rats that kill the young, and the golden toads of the Monte Verde cloud forest in Costa Rica appear to have been wiped out by a single El Niño. No doubt there have been many marine examples, but they have gone unrecorded; consequently, potential risks to marine organisms are rarely considered.

One option that is increasingly widely practiced for terrestrial species is *ex situ* conservation—the breeding of endangered species in zoological and botanical gardens and the storage of genetical material in banks of seeds or germplasm. Perhaps some effort should be made to develop the techniques for marine species along the lines that some algal species are maintained in cultures, in anticipation that the techniques will be needed once some prioritization is designated.

We remain totally uncertain as how best to respond to major events. The dramatic increase in abundance of the crown-of-thorns starfish (*Acanthaster planci*) elicited crisis responses with predictions of the catastrophic destruction of all coral reefs, whereas the immense oceanwide population outburst of the stomatopod *Oratosquilla investigatoris* in the western Indian Ocean in 1967 (Losse and Merrett 1971), which had a major impact on pelagic fisheries, went almost unnoticed. Should we continue to be worried by the reports of extensive bleaching of coral, or is this phenomenon another example of an event that lies within the range of tolerance of affected reefs?

One aspect about which there can be no doubt is the importance of limiting the accidental and purposeful introduction of exotic species. Although probably few introduced species do become established, those that do can create major environmental and resource management problems. The most obvious examples are the pests and diseases that accompany the translocation of commercial species, such as oysters. But fouling organisms create major problems for shipping and recreation, as seen in the introduction into UK waters of the barnacle *Chthalamus stellatus* and the brown seaweed *Sargassum muticum*. Estuarine species often move readily, and the Chinese mitten crab *Eriocheir sinensis* is causing problems to sea defenses in European waters. Most worrying is the possibility of the introduction of the species responsible for toxic algal blooms, which has led Australia to introduce restrictions on discharges of ballast water into coastal waters.

Diversity is regularly managed. Agricultural monocultures exemplify systems in which low diversity is maintained with total dominance by the crop plant. Conservation usually seeks to maximize both richness and evenness of communities, either by systematic cycles of perturbation (e.g., managing grazing or scrub clearance) or by habitat creation. Both are practiced to a very limited extent in the ocean—by establishing reserves, management of fish stocks, pollution controls, mariculture, and the creation of artificial reefs. Such management depends on the system being predictable and stable. Effective management of the open ocean can be achieved only if some of the sacred cows of international maritime law are laid to rest.

In unchanging environments, species richness and evenness increase with geological time to maxima. At low latitudes the greater climatic stability throughout the glacial cycles of the Quaternary has contributed to the latitudinal trend for species richness and evenness to be lower at high latitudes. As diversity increases, so does the complexity of the network of interactions tend to increase, reducing the system's resilience and slowing its recovery rate; it may become just as vulnerable to the accumulative effect of a series of frequent small perturbations as to a single large disturbance. These communities are fragile—readily changed irretrievably by disturbance and so potentially more vulnerable. Tropical rain forests, coral reefs, and abyssal benthic communities are all fragile and must be exploited with considerable care.

Not all ecosystems that are most at risk or vulnerable are so fragile. Estuaries are heavily exploited and polluted, but their communities are robust, naturally adapted to withstand wide fluctuations in their conditions. In robust systems, as long as the impact of human activity remains within the bounds of natural perturbations, the system will recover. However, there may be critical periods when natural variations are at their most extreme (e.g., during the seasonal cycle) when any additional "stress" erodes the system's ability to recover.

Deep Ocean Conservation

The vast size of oceanic systems has important implications for conservation and management of biodiversity. The creation of adequately sized reserves or parks would seem impractical; a park equivalent in size to the largest terrestrial reserve would be as inadequate as preserving a single square kilometer of the Serengeti Plain. The approach must be, therefore, to develop integrated systems of management for whole oceans. The present piecemeal approach as seen in the United Nations conventions on Marine Pollution (MARPOL) and the Law of the Sea (UNCLOS), the dumping conventions, and the fishery agreements are better than nothing but will ultimately prove inadequate.

Conservationists (e.g., McNeeley et al. 1990) have so far shown little interest in the deep ocean, assuming it is not under threat. Their focus on individual marine species is limited to whales, turtles, and some of the larger sharks and fishes. Broad concern has been expressed about the impact of "ghost fishing" by monofilament line and nets (particularly in the Pacific) and the general impact of oil spillages and litter. Commercial and sporting oceanic fishes have been subject to some resource regulation. However, management of fish stocks in coastal waters, despite considerable investment in research and data gathering, is only moderately effective. Coping with natural variability, managing multispecies stocks, and the socioeconomic aspects of regulating traditional activities in which there is a mismatch between the long-term economic investment and the shorter-term natural variations has proved to be intractable (e.g., McGlade and McGarvey 1992). In the Southern Ocean the effects of the drastic reductions, first of exploited fur seal populations and then of the large whales, on the structure and energy flow within the ecosystem has never been determined. Nor has research into assessing ecosystem responses to the removal of top predators by fishing been done; annually in the North Sea 65 to 70 percent of all the commercial-sized haddock are removed. What would the North Sea ecosystem be like without fishing?

The extent of this impact may be far greater than hitherto recognized. In New Zealand a small "no-take" reserve established at Leigh in the North Island has substantially improved local populations, even of migratory species. Ballantine (1991) argues forcibly that substantial areas of coastal waters (10 percent) should be set aside as nonexploited zones in order to conserve living resources. In Europe, resolving the confused mass of legislation to create such zones would be a major challenge.

Conservationists have argued that the maintenance of biodiversity is essential for the preservation of ecological processes (McNeeley et al. 1990). However, the link between biodiversity and process remains intuitive. The initial success of modeling in replicating events occurring at the onset of the Spring Bloom and the flows of carbon through the system, using relatively simple flow models (Fasham et al. 1990), suggests that the degree of complexity of an ecosystem may bear little relationship to its overall performance. Processes do not seem to be tightly coupled with the composition of the system. This would have far-reaching implications for both the management of oceanic ecosystems and the way in which they are eventually exploited.

In trying to establish conservation priorities, emphasis has been placed on maximizing the numbers of species conserved and hence on the need to establish more accurate listings of what species occur where (Hill 1991). The success of critical faunal analyses (e.g., Vane-Wright et al. 1991) depend on the existence of substantial and well-documented knowledge of the fauna and flora. Although this approach includes assessments of systematic differences at family and generic level, it still tends to place maximum emphasis on tropical regions. The question then becomes whether the emphasis should be solely on these aspects of species richness or whether the systems and processes that support global diversity should be included. If the former, then the total conservation effort will be concentrated at low latitudes, whereas many of the biological processes that maintain global health occur at middle to high latitudes. Following the logic of emphasizing species richness in pelagic oceanic systems would lead to conservation efforts being focused on depths of 500 to 1,000 meters, ignoring the euphotic zone whose processes support the richness (e.g., figure 11).

The scales of the processes determining the distributions of many oceanic species are considerably larger than those on land. Conservation of oceanic ecosystems and processes then becomes a rather different problem from the conservation of terrestrial species.

How is progress to be advanced?

The block to progress is ignorance. Some of this ignorance is unnecessary. There are no inventories of known species, and biological data have never been included in international oceanic data bases. Large historical collections of pelagic samples, which could be analyzed and provide much improved coverage of a wide range of taxa, are under threat of being discarded. Unlike long-term sampling programs, zoogeography has not had the renaissance it deserves; and yet without good spatial coverage (Angel 1991), adequate interpretation of data in the time realm is compromised.

Other areas of ignorance are not so easy to resolve. How does organization and structure at one level of scale and complexity influence those at both higher and lower levels? How do taxa with totally differing time/space scales interact within the ecosystems? (A generation time for a whale is 10^3 to 10^4 times that for bacteria and phytoplankton.) Oceans are nonlinear, so influences cascade up and down scales of variability; similar cascading can be expected between the different organizational levels of assemblages and ecosystems. We are only just beginning to be able to describe oceanic diversity and disparity. Yet, even though we are far from being able to understand its importance, judgments have to be made *now* about future policies. Application of the precautionary principle, whereby decision makers need to either demonstrate the environmental safety of their policies or assume the very worst case scenario, runs the danger of creating inaction. However, until ecological considerations are given improved priority relative to short-term economic considerations, there is no other prudent course to take.

As UNEP moves on in the processes surrounding a Convention on Global Biodiversity, we will be no closer to providing rational and well-supported arguments for the conservation of biodiversity, particularly in the oceans, unless important defects are corrected. The oceanographic community must act, using whatever information is available to enlighten the politicians as to how oceanic systems differ from terrestrial systems; otherwise, we will be faced with international legislation that may prove to be quite inappropriate for most of the earth's surface.

Conclusions

1. Emphasis on diversity to the exclusion of disparity will lead to the oceans receiving little if any priority in conservation and resource management. The relative importance of ocean processes vis-à-vis terrestrial processes to global ecology indicates that this imbalance could lead to serious errors in decision making.

2. The scale of oceanic systems is so large that the methodology adopted for terrestrial conservation and resource management can not be directly applied. To be effective, protected areas would need to be the size of the ocean gyres, which is unlikely to be achievable. Instead, integrated management protocols will need to be developed.

3. The theory developed to explain diversity in terrestrial ecosystems is demonstrably inadequate to explain diversity in oceanic pelagic communities.

4. Integrated management protocols must be based on the best possible science. The most obvious gaps are (a) a lack of a comprehensive inventory of known species and their zoogeographical distributions and (b) the missing link between biodiversity, which is being used as the basis for present management, and the ecological processes in the ocean, which play such an important role in global homeostasis.

5. The information base on which to develop a predictive understanding of the interaction between diversity and ecological process can be greatly enhanced relatively inexpensively by systematically collating existing data and working up extant collections of material.

References

Angel, M.V. 1991. Variations in time and space: Is biogeography relevant to studies of long-time scale change? *Journal of the Marine Biological Association of the United Kingdom* 71:191–206.

Angel, M.V., and A. de C. Baker. 1982. Vertical standing crop of plankton and micronekton at three stations in the Northeast Atlantic. *Biological Oceanography* 2:1–30.

Ballantine, W. 1991. *Marine reserves for New Zealand.* University of Auckland, Leigh Laboratory Bulletin, no. 25, 1–196.

Barron, J.A., and Baldauf, J.G. 1989. Tertiary cooling steps and paleoproductivity as reflected by diatoms and biosiliceous sediments. In *Productivity of the ocean: Present and past,* ed. W.H. Berger, V.S. Smetachek, and G. Wefer, 341–354. New York: John Wiley & Sons Ltd.

Cline, R.M., and Hayes, J.D., eds. 1976. Investigation of late quaternary paleoceanography and paleoclimatatology. *Memoirs of the Geological Society of America* 145:1–464.

Dickson, R.R., J. Meincke, S.A. Malmberg, and A.J. Lee. 1988. The "Great Salinity Anomaly" in the northern North Atlantic. *Progress in Oceanography* 20:103–151.

Elton, C.S. 1966. *The pattern of animal communities.* London: Methuen & Co. 432 pp.

Fasham, M.J.R., H.W. Ducklow, and D.S. McKelvie. 1990. A nitrogen-based model of plankton dynamics in the oceanic mixed layer. *Journal of Marine Research* 48:591–639.

Fleminger, A. 1986. The Pleistocene equatorial barrier between the Indian and Pacific Oceans and a likely cause for Wallace's Line. In *Pelagic Biogeography,* UNESCO Technical Papers in Marine Science 49:84–97.

Greig-Smith, P. 1964. *Quantitative plant ecology.* London: Butterworth's. 256 pp.

Haedrich, R.L., and N.R. Merrett. 1990. Little evidence for faunal zonation or communities in deep-sea demersal fish faunas. *Progress in Oceanography* 24:239–250.

Haq, B.U. 1984. Paleoceanography: A synoptic overview of 200 million years of ocean history. In *Marine geology and oceanography of Arabian Sea and coastal Pakistan,* ed.

B.U. Haq and J.D. Milliman, 201–232. New York: Van Nostrand Reinhold.

Haury, L.R., J.A. McGowan, and P.H. Wiebe. 1978. Patterns and processes in the time-space scales of plankton distribution. In *Spatial patterns in plankton communities,* ed. J.H. Steele, 277–327. New York: Plenum.

Hill, J. 1991. Conserving the world's biological diversity; How can Britain contribute? Proceedings of a seminar presented by the Department of the Environment in association with the Natural History Museum, June 1991, Department of the Environment, UK. 83 pp.

Jaanusson, V. 1981. Functional thresholds in evolutionary progress. *Lethaia* 14:251–260.

Lamb, H.H., 1977. *Climate: Present, past and future.* Vol. 2, *Climatic history and the future.* London: Methuen & Co. Ltd. 935 pp.

Losse, G.F., and N.R. Merrett. 1971. The occurrence of Oratosquilla investigatoris (Crustacea: Stomatopoda) in the pelagic zone of the Gulf of Aden and the equatorial western Indian Ocean. *Marine Biology* 10:244–253.

MacArthur, R.H., and E.O. Wilson. 1967. *The theory of island biogeography.* Princeton, N.J.: Princeton University Press. 203 pp.

McGlade, J.M., and R. McGarvey. 1992. Integrated fisheries management models: Understanding the limits to marine resource exploitation. In *Advances in the science and technology of ocean management,* ed. H. Smith, 194–232. London and New York: Routledge.

McGowan, J.A. 1990. Species dominance-diversity patterns in oceanic communities. In *The earth in transition,* ed. G. M. Woodwell, 395–421. New York: Cambridge University Press.

McGowan, J.A., and P.W. Walker. 1985. Dominance and diversity maintenance in an oceanic ecosystem. *Ecological Monographs* 55:103–118.

McNeeley, J.A., K.R. Miller, W.V. Reid, R.A. Mittermeier, and T.B. Werner. 1990. *Conserving the world's biological diversity.* Washington, D.C.: International Union for Conservation of Nature and Natural Resources (IUCN-Gland, Switzerland), World Resources Institute (WRI), Conservation International (CI), World Wildlife Fund-United States (WWF-US), and the World Bank. 193 pp.

Platt, T., K.H. Mann, and R.E. Ulanowicz, eds. 1981. *Mathematical models in biological oceanography.* Paris: UNESCO Press. 156 pp.

Ray, G.C. 1991. Coastal-zone biodiversity patterns. *BioScience* 41:490–498.

Runnegar, B. 1987. Rates and models of evolution in Mollusca. In *Rates of evolution,* ed. K.S.W. Campbell and M.F. Day, 39–60. London: Allen & Unwin.

Sournia, A. Chretiennot-Dinet, and M. Ricard. 1991. Marine phytoplankton: How many species in the world ocean? *Journal of Plankton Research* 12:1093–1099.

van der Spoel, S., and A.C. Peirrot-Bults, eds. 1979. *Zoogeography and diversity of plankton.* Utrecht: Bunge.

van der Spoel, S., A.C. Pierrot-Bults, and P.H. Schalk. 1990. Probable mesozoic vicariance in the biogeography of Euphausiacea. *Bijdragen tot de Dierkunde* 60:155–162.

Tunnicliffe, V. 1991. The biology of hydrothermal vents: Ecology and evolution. *Oceanography and Marine Biology Annual Review* 29:319–407.

Turner, R. 1981. Wood islands and thermal vents as centres of diverse communities in the deep sea. *Soviet Journal of Marine Biology* 7:1–10.

Vane-Wright, R.I., C.J. Humphries, and P.H. Williams. 1991. What to protect? Systematics and the agony of choice. *Biological Conservation* 55:235–254.

Venrick, E.L. 1986. Patchiness and the paradox of plankton. UNESCO Technical Papers in Marine Science 49:261–265.

Whitmore, T.C. 1982. On pattern and process in forest. In *The planet community as a working mechanism,* ed. E. Newman, 45–59. Special Publications of the British Ecology Society, 1. Oxford: Blackwell.

3
Oceanic Species Diversity
John A. McGowan

Because of large-scale and accelerating rates of extinction of species, it has become essential that we develop the knowledge necessary to assess the consequences of these losses. There is some evidence from a few systems that the exclusion of certain species (called *key species*) from communities results in a rearrangement of the way in which abundances are partitioned among species and additional changes in diversity. Further, there is much speculation that the loss of any species that is a normal and regular part of a community will also have disruptive effects. These effects are thought to come about because of the close coupling between ecosystem species structure and system function. System function includes the rate of processing atmospheric gases such as oxygen (O_2) and carbon dioxide (CO_2), the cycling of essential plant nutrients such as nitrogen and phosphorus, and the sequestering of pollutants such as radioisotopes and manufactured organic compounds. Further, system primary productivity, the efficiency with which energy is transferred through the food web, and even the fraction allocated to decomposers may depend on the maintenance of the diversity of communities. In such systems we expect (1) that energy and materials flow systematically through pathways that have developed over a very long period of natural selection and (2) that competitive forces have tended to maximize the efficiency with which each entity or link (=species) within the community food web processes resources. These entities achieve this efficiency by specializing on certain resources. If this "competitive equilibrium" or "resource allocation" version of nature is essentially correct, then we may expect that the role of each link is important.

The removal of one or more links can lead to a rearrangement of the structure of pathways and a loss of efficiency, usually for an unknown period of time and to an unknown degree. The entire functioning of the system as a pool or reservoir or productive organization would differ because the turnover times and pathways through the machinery would change. A loss of many species might be very disruptive. In addition to the potential disruption of the function of pelagic systems, the simple loss of genetic material, much of it unique to the open ocean, is an event of unknown seriousness. Science has only recently recognized the vast potential—the genetic storehouse—in wild populations, and we can predict that in the next 50 to 100 years the utilization of this treasure will be realized. We are already doing so with land plants and animals. This work has seen a virtual exponential increase, and there is every reason to believe that the ocean will soon be exploited in a similar way. The pelagic realm represents the last and the greatest frontier in this quest.

Further, we do not really understand what is responsible for the origin of the diversity patterns we see in the open ocean or what maintains them. Evolutionary theory, created to explain land patterns, does not satisfactorily explain the observations made in the open ocean. But because the oceans are so large and oceanographers so few, even our observations (i.e., descriptions) of the pelagic realm are for the most part vague and uncertain.

What is needed now are inventories of pelagic species diversity patterns, their gradients, thresholds, and boundaries and how these relate to the currents, upwelling, downwelling, and great natural perturbations such as El Niños. Although some of this sort of work has been done in the past (United States, United Kingdom, Japan, the former USSR), it has not had recent support. Curiously, most of the necessary collections have already been made on the great oceanic expeditions of the 1960s and 1970s. What is lacking is support for the study of this material from the viewpoint of determining diversity patterns. Because this work was not fashionable for the past decade and a half, the expertise necessary to do the work has not been passed on to a new generation. This must change if we are to pursue this effort.

Recommendations

1. Training grants for taxonomic studies of pelagic organisms.
2. Endowed positions at museums, universities and, above all, oceanographic institutions for biogeographic research.
3. Support for biogeographic studies for (the few) existing specialists.
4. Support for further open-ocean collections for biogeographic purposes.
5. Support for symposia and meetings on pelagic diversity problems.

4
Biodiversity: Human Impacts through Fisheries and Transportation

Makoto Omori, Christopher P. Norman, and Hiroshi Yamakawa

Biological diversity can be defined on several biological levels: ecosystem level, species level within that ecosystem, and genetic level within species themselves (Soule 1991). An alternative assessment of diversity is that of ecological diversity (Ray 1986), i.e., the diversity of the life forms rather than individual species themselves; in this sense, grouping is functional and not necessarily related to species taxonomy. The diversity most often referred to, however, is species diversity. Species diversity is a concise expression of how many species live in a community (*richness*) and how these species are distributed within that community (*evenness*). Previous literature suggests that species diversity is related to the degree of complexity of the flow of energy within a community and its environmental characteristics (Paine 1966, Menge and Sutherland 1976). Communities with high diversities tend to occur in areas of environmental stability (or predictability). Species diversity is generally greater in the waters of lower latitudes than in those of higher latitudes, is commonly greater in marine environments than in freshwater ones, and is lowest in brackish water regions (Dumbar 1960, Connell 1978). Pollution lowers species diversity, and therefore the temporal variation of diversity has been used in assessing the impact of pollutants on marine flora and fauna (Patten 1962).

At present, the sea faces an unprecedented threat to its biological diversity due to the increasing chronic stresses caused by pollution and exploitation of biological and mineral resources. Among the human impacts, pollution, fisheries, and transportation seems to bring a most visible effect on marine biological diversity over a relatively short time. For example, from discovery to extinction of Steller's sea cow took only

27 years. Similarly, the spread of dinoflagellates causing red tides and/or shellfish poisoning appears to have been greatly accelerated by modern shipping methods in combination with the chronic levels of eutrophication in many coastal areas of the world.

According to the United Nations' Food and Agriculture Organization (FAO), the global catch of fishes totals about 99 million metric tons at present (FAO 1991). By the year 2000, FAO estimates that another 19 million metric tons will be needed simply to maintain consumption at the current annual rate of about 12 kilograms per capita. To enhance productivity to maintain per capita consumption levels, new resources and wise utilization and management of existing stocks as well as new culture and enhancement techniques will be increasingly necessary. In this paper we focus on fishing, on transport, and on enhancement fisheries as human impacts on diversity of the seas and review their potential threats and any possible countermeasures to protect what ultimately may be one of the most important resources of all, biological diversity itself.

Fishing Activity

The effect of fishing on biological diversity and ecosystem structure has until recently not been clearly appreciated. Difficulties arise in distinguishing changes caused by natural phenomena and those caused by fishing. Examples of such perturbations are the succession of species in fishing stock, and this theme has been reviewed by a number of scientists such as Hempel (1978) and Sherman (1988). Evidence from these reviews suggests that marine ecosystems are very dynamic, with their biological components and structure fluctuating with environmental and fishing pressures. The driving forces suggested for changes in biological diversity within a marine ecosystem have been both fisheries and environmental factors and frequently a combination of the two.

Early in this century, the fishery in the North Sea saw a decline of the herring and mackerel stocks along with their principal predator, the bluefin tuna, and a concurrent increase in stocks of gadoids (Holden 1978). A simulation model by Anderson and Ursin (1978) found the larval mortality rate to be a key factor in the success of recruiting to the adult population. When herring and mackerel stocks were abundant, these pelagic species preyed heavily on the larvae of other species such as the gadoids. With the overexploitation of the herring and mackerel stocks by humans, this predation pressure on gadoid larvae was removed and survival in the juvenile stages of gadoids improved, thus leading to greater recruitment. Adult gadoids further prey on herring and juvenile mackerel, thus maintaining the new ecological balance. How humankind has influenced this change remains unclear; however, the change of the key species of that ecosystem from a pelagic predator to a demersal predator must have had profound effects on the structure of the ecosystem of the North Sea as a whole. Temperate fisheries appear to affect predominantly the species evenness within an ecosystem; in mixed fisheries, however, where fishing effort and return may have less direct relationship, species richness may also be at risk.

Tang (1987) reports the effects of overfishing and pollution in the Yellow Sea. Between 1960 and 1980 there was a 40 percent decrease in biomass of commercially important demersal stocks, represented by the yellow croaker (*Pseudosciaena spp.*), as evidenced by a marked reduction in mean body size and a concurrent increase in biomass of low-value, pelagic species such as anchovy (*Setipinna taty*). Similar to the North Sea fishery, the Yellow Sea fishery has shown marked changes in the quantity and quality of catch and ultimately in the ecosystem structure. Due to the decline in traditional stocks, artificial enhancement programs have been developed; with demersal predator pressure reduced, programs to enhance the prawn (*Penaeus orientalis*) fishery have shown remarkable success. In Japan, enhancement programs are also commonly used for marine and freshwater species. (The section "Enhancement Fisheries," below, details problems associated with this practice.)

Inadvertent Transport of Microorganisms

Attachment to the hulls of vessels has long been recognized as a mechanism that can carry living marine organisms. In recent years, however, the chance of dispersal of marine organisms into new environs far removed from their previously known distribution has certainly been increasing. Although other possible explanations for the apparent increased distributional area are dispersion by ocean circulation or previous lack of adequate faunal and floral investigation in the study area, it is more reasonable and likely in many cases that the cause of dispersion is by human activities.

The transport of coastal species through ballast water has become more significant with the development of antifouling paint, the application of steel in shipbuilding, and the increase of speed and distance traveled (Hutchings et al. 1987). Large cargo vessels usually take in up to 100,000 tons of ballast water while in port, and the possibility of transoceanic and interoceanic dispersal of phytoplankton cysts and zooplankton, including larvae of marine invertebrates, has been increased particularly by ballast water transport (Brylinski 1981, Carlton 1985, Williams et al. 1988, Hallegraeff and Bolch 1991). Medcof (1975) found numerous living planktonic crustaceans and polychaete larvae in the ballast water carried from Japan to New South Wales, Australia (a distance of nearly 9,000 kilometers). According to Hallegraeff and Bolch (1991), of 80 cargo vessels entering Australian ports, 40 percent contained viable dinoflagellate cysts and 6 percent carried the cysts of the toxic species *Alexandrium catenella* and *A. tamarense*.

The development of aquaculture has created another possibility for the dispersal of marine organisms: organisms can be passively carried in the water of the container transporting the species for introduction. The recent finding of an Asian estuarine copepod (*Pseudodiaptomus marinus*) in coastal embayments of southern California is most likely a direct consequence of an aquaculture project (Fleminger and Hendrix Kramer 1988). According to these authors, there is a possibility that colonization of an embayment by *P. marinus* has somehow resulted in the exclusion of the local species *P. euryhalinus*. It is noteworthy that disjunct populations of the North Pacific calanoids *Centropages abdominalis* and *Acartia omorii* and the East Asian coastal oithonid *Oithona davisae* have been found in a south Chilean fjord that has extensive aquaculture programs utilizing Japanese salmon (Hirakawa 1986, 1988).

Enhancement Fisheries

In the recent past, human beings have turned from being merely hunters/gatherers of the seas to more fully utilizing the seas' resources by farming. Worldwide, the scale and application of aquacultural techniques is probably most advanced in Japan. Of the total Japanese offshore and coastal fisheries product of 8,463,000 metric tons, culture practices account for 1,272,000 metric tons, or approximately 15 percent of the total (Norin Tokei Kyokai 1991).

In Japan, aquaculture of fish and invertebrates falls into two categories. The first is represented by *cage culture techniques,* where food is given and the resource is maintained at one location. Species for which this style of culture is particularly important are salmon (*Oncorhynchus kisutch*), yellowtail (*Seriola quinqueradiata*), and the red sea bream (*Pagurus major*). The second method is *mariculture*, in which recruitment is enhanced by rearing the larval and juvenile stages of desired species in, as near as possible, ideal environmental conditions, free from starvation and predation, thus allowing a high survival rate. These are then released into the natural environment to be later harvested via fishing techniques. With the reduction of the catch of some commercial species, this mariculture program is of considerable financial, social, and political importance in Japan. Current production levels of juveniles of some representative species are given in table 1.

One of the most disturbing problems of mariculture concerns the genetic identity of released juveniles. Within sexually reproducing species there is a degree of genetic diversity exhibited by all individuals. Humans have defined *species* as a group of individuals in which interbreeding is possible, giving rise to reproductively viable offspring. This definition, however, obscures many of the genetic complexities within a species, with many species showing populations with morphological, reproductive, and behavioral differences. The identity of populations has been most clearly established for commercially important species such as the herring (Iles and Sinclair

Table 1
Seeding production in Japan of some representative marine species, 1989

	(in thousands of individuals)
Salmon (*Oncorhynchus keta*)	2,060,000
Red sea bream (*Pagurus major*)	77,569
Black sea bream (*Acanthopagrus schlegelii*)	9,037
Flatfish (*Limanda yokohamae*)	1,480
Flounder (*Paralichthys olivaceus*)	28,331
Prawn (*Penaeus japonicus*)	523,695
(*Metapenaeus monoceros*)	40,862
Crab (*Portunus trituberculatus*)	52,459
Scallop (*Patinopecten yessoensis*)	3,311,609
Abalone (*Nordotis spp.*)	21,133
Sea urchin (*Strongylocentrotus intermedius*)	23,067

Source: Japan Fishery Agency.

1982) and salmon (Palva et al. 1989). These populations have evolved over geological time, giving rise to populations suited to their respective environments. The preservation of this genetic integrity is of fundamental ecological importance if humans are to maintain this genetic diversity for future use. Hatchery parental brood stock are often of various genetic backgrounds, either inadvertently or deliberately, giving rise to hatchery fingerlings whose genetic makeup is different from that of the indigenous population. Nonindigenous hatchery fish mix and genetically pollute the indigenous population, which lose their genetic identity; and, because they have evolved to fit their ecological niche, the new genetic material undermines the adaptations of the resident stock. Clear case studies in which unique native stock have been threatened from hatchery releases are given by Garcia-Marin et al. (1991) and Krueger and May (1991).

Apart from genetic pollution by released fish, behavioral differences between native and released fish also cause problems. In Japan, the release into rivers of ayu (*Plecoglossus altivelis*), a herbivorous annual anadromous fish of considerable sporting and economic value, has caused genetic change in wild populations (Taniguchi et al. 1983.) In this species, a favored stocking fish is a nonanadromous variety from Lake Biwa in central Japan that has stronger territorial behavior than, and thus causes habitat loss to, its wild counterparts. Ayu of anadromous stock (of various origin) reared by interbreeding have further altered the genetic nature of the indigenous populations, which Taniguchi et al. (1983) estimated to have required more than 15,000 years to differentiate. This release process has been carried out over a period of up to 80 years in many Japanese rivers and has caused a significant loss of genetic variability and a lowered survival rate of indigenous fish, thus causing an increased dependence on hatchery-produced fish. Similar effects have been suggested for populations of the red sea bream (Taniguchi 1986), and morphological abnormalities of adult fish have recently been observed in some stocking areas around the Japanese coast (H. Yamakawa, personal observation).

Added concern arises where large numbers of trophically similar consumers are released into a small geographical area, leading to competition with wild stock for food resources. For example, the number of planktivorous juveniles of the red sea bream and dog salmon released in Japan reaches 77×10^6 and $2,060 \times 10^6$ individuals per year, respectively (table 1). These large numbers of juveniles will, if they are to survive, compete for food resources. The carrying capacity, dispersion rate, and density-dependent factors have not been examined for many species released in Japan, nor have the effects of such a release on the other biological components of the ecosystem. Concern particularly occurs where food resources are limited, leading to trophic competition between wild and stocked individuals and, where densities are high, possible failure of both due to malnutrition.

The opportunities to improve maricultural techniques will be greatly enhanced by biotechnological advancement, but with this goes the responsibility of protecting and managing the natural resources. Particularly, the release of alien species and the use of triploid and transgenic fish should be carefully assessed in relation to preservation of natural genetic material (Hallerman et al. 1991; Kapuscinski and Hallerman 1991; Seto 1991).

Discussion

Considering biological diversity, there are some fundamental differences between marine and terrestrial systems. First, species diversity in terrestrial systems is generally supposed to be much higher than in marine systems; at a more basic taxonomic level such as phyla, however, diversity of marine fauna is much greater than that of terrestrial fauna (Ray 1986). On a genetic level, we do not as yet know which system shows higher diversity. Regarding rare marine species, except for very localized, endemic species, the assessment and monitoring of rare species is particularly difficult due to the sheer size of the marine habitat.

Second, marine food webs and trophic relationships are different from terrestrial ones, as the structure and dynamics of food webs are governed by the size of the individual rather than by species. For example, the eggs and newly hatched fish larvae are ready food items for carnivorous copepods; but with larval development the predator/prey relationship is switched, with the copepods ultimately becoming an important trophic resource for many fish species.

Third, the terrestrial environment has largely been divided by humans into expanses of wild and largely underexploited areas and into greatly altered agricultural areas in which, by human manipulation, the diversity has been greatly reduced. In marine systems, by and large, humans are still exploiting truly wild populations, often with minimal management. For the management of marine biodiversity these intrinsic differences between terrestrial and marine environments should be taken into consideration.

The effect of fishing activity on the marine ecosystem has long been a subject of debate. Humans, as very effective predators, appear to have the most intense effects on ecosystems where the environmental conditions are stable. Fluctuations in environmentally driven ecosystems have also been shown to have dramatic effects on marine and associated coastal ecosystems, e.g., as seen during El Niño years. Clear evaluation of the effect of fishing on diversity is thus often masked by environmental perturbations and is difficult to assess due to the vastness and three-dimensional aspect of the ocean as well as the lack of adequate previous investigation. How humans are affecting diversity through fishing may be most readily assessed from ecosystems with stable or at least predictable environmental conditions. The Antarctic marine realm—as a largely pollution-free, environmentally predictable, simple ecosystem with well-described biological components including indicator and keystone species—might particularly lend itself to theoretical evaluation of fishery pressure on biodiversity.

Recent advances in marine transportation appear to have affected coastal planktonic and benthic communities markedly by introducing harmful toxic dinoflagellates into many areas of the world (Hallegraeff et al. 1990, Smayda 1990). For many locations, however, whether there has been continuous transport of these species but inappropriate conditions for propagation or whether these toxic flagellates are indigenous but have merely been unobserved remains unknown (Smayda 1990). Studies by Hallegraeff et al. (1990) and by Hallegraeff and Bolch (1991) have clearly ascertained that dinoflagellates or their cysts, including *Gymnodinium catenatum*, which can cause paralytic shellfish poisoning, were introduced into Tasmanian coastal waters by ships. Considering the volume of world shipping during the last decades, it is suggested that one or more factors other than transport, such as a degree of eutrophication, are necessary to allow *G. catenatum* to establish itself in new environs. The lack of adequate previous investigation of fauna and flora in study areas makes analysis of environmental versus transportation-of-species factors unclear. To assess the effect of transport of alien species as well as pollution and how the two may be interconnected, taxonomic and systematic studies need to be supported. The current shortage of taxonomists has resulted in a lack of clear data on which to assess such changes in biota.

How we enhance fisheries and maintain genetic variability obviously requires careful attention if the genetic information stored in unique populations is to be maintained for future use. Much of the basic research into genetics of fish stocks has not yet been thoroughly carried out, with the result that little is known about the stock identity of many species (Japan Fisheries Resource Conservation Association 1989). The appreciation, conservation, and future usage of the sea's genetic resources rely on basic research to clarify what resources are available. Clearly, some code of practice is required to minimize the effects of inconsistent stock, with particular responsibility falling on hatchery staff to realize the implications of the release of fish of nonconsistent genetic stock and especially alien species and genetically manipulated stock.

On writing this paper, we have found very few well-documented accounts of how humans are affecting marine biological diversity. That marine diversity is affected is abundantly clear, as fishermen and people whose work is connected to the sea will readily confirm. Documenting this process, compared to terrestrial environments, is complex, however, due to the very nature of the sea. It is worth bearing in mind that terrestrial ecological theory may be inadequate to explain processes controlling marine biodiversity. Changes of biological diversity with time are inherently natural and dynamic, but humans are increasingly influencing this balance. Human population increase will progressively threaten the biological diversity of coastal areas, particularly in developing and underdeveloped countries. We should not forget that productivity is what interests most of humanity and that the altruistic ideal of preserving marine diversity may not be universally accepted.

References

Anderson, K.P., and E. Ursin. 1978. A multispecies analysis of the effects of variations of effort upon stock composition of eleven North Sea fish species. *Papp. P.-v. Reun. Cons. int. Explor. Mer.* 172:286–291.

Brylinski, J.M. 1981. Report on the presence of *Acartia tonsa* Dana (Copepoda) in the Harbour of Dunkirk (France) and its geographical distribution in Europe. *Journal of Plankton Research* 3:255–260.

Carlton, H.T. 1985. Transoceanic and interoceanic dispersal of coastal marine organisms: The biology of ballast water. *Oceanography and Marine Biology Annual Review* 23: 313–371.

Connell, R.J. 1978. Diversity in tropical rain forests and coral reefs. *Science* 199:1302–1310.

Dumbar, M.J. 1960. The evolution of stability in marine environments. Natural selection at the level of the ecosystem. *American Naturalist* 94:129–136.

Fleminger, A., and S. Hendrix Kramer. 1988. Recent introduction of an Asian estuarian copepod, *Pseudodiaptomus marinus* (Copepoda:Calanoida), into southern California embayments. *Mar. Biol.* 98:535–541.

Food and Agriculture Organization (FAO) of the United Nations. 1991. Fisheries statistics—catches and landings. Vol. 68, FAO Fisheries Series (36). 517 pp.

Garcia-Marin, J.L., P.E. Jorde, N. Ryman, F. Utter, and C. Pla. 1991. Management implications of genetic differentiation between native and hatchery populations of brown trout (*Salmo trutta*) in Spain. *Aquaculture* 95:235–249.

Hallegraeff, G.M., and C.J. Bolch. 1991. Transport of toxic dinoflagellate cysts via ship's ballast water. *Marine Pollution Bulletin* 22:27–30.

Hallegraeff, G.M., C.J. Bolch, J. Bryan, and B. Koerbin. 1990. Microalgal spores in ship's ballast water: A danger to aquaculture. In *Toxic marine phytoplankton*, ed. E. Graneli et al., 475–480. New York: Elsevier.

Hallerman, E.M., A.R. Kapuscinski, P.B. Hachett, A.J. Faras, and K.S. Guise. 1991. Gene transfer in fish. In *Advances in fisheries technology and biotechnology for increased profitability,* ed. M.N. Voigt and J.R. Botta, 35–49. Lancaster: Technomic Publishing.

Hempel, G. 1978. Symposium on north sea fish stocks—recent changes and their causes. *Rapp. P.-v. Reun. Cons. int. Explor. Mer.* 172:5–9.

Hirakawa, K. 1986. A new record of the planktonic copepod *Centropages abdominalis* (Copepoda, Calanoida) from Patagonian waters, southern Chile. *Crustaceana* 51:296–299.

_____. 1988. New records of the North Pacific coastal planktonic copepods, *Acartia omorii* (Acartiidae) and *Oithona davisae* (Oithonidae) from southern Chile. *Bulletin of Marine Science* 42:337–339.

Holden, M.J. 1978. Long-term changes in landings of fish from the North Sea. *Rapp. P.-v. Reun. Cons. int. Explor. Mer.* 172:11–26.

Hutchings, P.A., J.T. van der Velde, and S.J. Keable. 1987. Guidelines for the conduct of surveys for detecting introductions of non-indigenous marine species by ballast water and other vectors—a review of marine introductions to Australia. *Occ. Rep. Australian Mus.* 3:1–147.

Iles, T.D., and M. Sinclair. 1982. Atlantic herring: Stock discreteness and abundance. *Science* 215:627–633.

Japan Fisheries Resource Conservation Association. 1989. *Isozyme analysis of fish and invertebrates* (in Japanese). Special Report of the Working Group for the Development for New Fishery Technology. Tokyo. 555 pp.

Kapuscinski, A.R., and E.M. Hallerman. 1991. Implications of introduction of transgenic fish into natural ecosystems. *Can. J. Aquat. Sci.* 48:99–107.

Krueger, C.C., and B. May. 1991. Ecological and genetic effects of salmonid introductions in North America. *Can. J. Fish. Aquat. Sci.* 48:66-77.

Medcof, J.C. 1975. Living marine animals in a ship's ballast water. *Proc. Natl. Shellfish Assoc.* 65:54–55.

Menge, B.A., and J.P. Sutherland. 1976. Species diversity gradients: Synthesis of the roles of predation, competition, and temporal heterogeneity. *American Naturalist* 100:351–369.

Norin Tokei Kyokai. 1991. White paper on fisheries, 1990 (in Japanese) (Gyogyo Hakusho). Tokyo. 274 pp.

Paine, R.T. 1966. Food web complexity and species diversity. *American Naturalist* 100:65–75.

Patten, B.C. 1962. Species diversity in net phytoplankton of Raritan Bay. *Journal of Marine Research* 20:57–75.

Palva, T.K., H. Lehvaslaiho, and E.T. Palva. 1989. Identification of anadromous and non-anadromous salmon stocks in

Finland by mitochondrial DNA analysis. *Aquaculture* 81:237–244.

Ray, G.C. 1986. Ecological diversity in coastal zones and oceans. In *Biodiversity*, ed. E.O. Wilson, 36–50. Washington, D.C.: National Academy Press.

Seto, A. 1991. Overview of marine biotechnology in Japan and its commercial application for aquaculture. In *Advances in fisheries technology and biotechnology for increased profitability*, ed. M.N. Voigt and J.R. Botta, 525–541. Lancaster: Technomic Publishing.

Sherman, K. 1988. Large marine ecosystems as global units for recruitment experiments. In *Toward a theory on biological-physical interactions in the world oceans*, ed. B.J. Rothschild, 459–476. Kluwer Academic Publishers.

Smayda, T.J. 1990. Novel and nuisance phytoplankton blooms in the sea: Evidence for a global epidemic. In *Toxic marine phytoplankton*, ed. E. Graneli et al., 29–40. New York: Elsevier.

Soule, M.E. 1991. Conservation: Tactics for a constant crisis. *Science* 253:744–750.

Tang, Q., 1987. Changes in biomass of the Huanghai Sea ecosystem. In *Biomass and geography of large marine ecosystems*, ed. K. Sherman and L.M. Alexander, 7–35. Boulder, Colo.: Westview Press.

Taniguchi, N., S. Seki, and Y. Inada. 1983. Genetic variability and differentiation of amphidromous, landlocked, and hatchery populations of ayu *Plecoglossus altivelis* (in Japanese). *Bull. Japan Soc. Sci. Fish.* 49:1655–1663.

Taniguchi, N. 1986. Genetic problems in seed production (in Japanese). In *Sea farming technology of red sea bream*, ed. M. Tanaka and Y. Matsumiya, 37–58. Tokyo: Koseisha-Koseikaku.

Williams, R.J., F.B. Griffiths, E.J. van der Wall, and J. Kelly. 1988. Cargo vessel ballast water as a vector for the transport of non-indigenous marine species. *Est. Coast. Mar. Sci.* 26:409–420.

5
Mariculture on Coastal China and Biodiversity
C. K. Tseng

Coastal China has been engaged in fishing and marine harvesting activities for ages, and we have depended for most of this time on the natural stock-replenishing ability of the fish and related products. Natural replenishment had been sufficient in most cases when our fishing methods were primitive and the rate of natural replenishment exceeded that of depletion by fishing. In recent years, however, as fishing methods became more modern and advanced, and the rate of depletion of natural resources exceeded that of natural replenishment, overfishing resulted and certain kinds of fish, shrimp, mollusks, seaweeds, and other natural populations under excessive exploitation gradually diminished in stock until their supply was unable to meet the demand. Mariculture of the overexploited organisms was therefore undertaken.

For quite some years there has been a controversy regarding mariculture and biodiversity. Some insist that mariculture is detrimental to the environment and biodiversity, and others believe that mariculture is beneficial to the environment and biodiversity. A historical review of progress in mariculture in the People's Republic of China will demonstrate the great antiquity of the practice in China and its recent important expansion. It will also serve as a basis to examine aspects of this issue.

The Historical Perspective

Mariculture may be defined as *the science and art of commercial cultivation of marine organisms* and is a branch of aquaculture dealing with the commercial cultivation of aquatic organisms in general.

Mariculture in China is both old and new. It is old, because mariculture started in China about one thousand years ago with the cultivation of the glueweed *Gloiopeltis* and the oyster *Crassostrea*. The glueweed has been cultivated in Jinmen County (Quimoy) near Xiamen (Amoy) in southern Fujian Province since the Song dynasty (A.D. 960–1279); using a simple rock-cleaning method, the rocks bearing the glueweed are cleaned in spring to let the glueweed's spores settle and grow. In Jinmen the littoral rocks in many places were privately owned; the glueweed was primarily used in the textile industry. By the same method the *zicai*, or purple laver, was cultivated in Pingtan County (Haitan), also of Fujian Province, more than two hundred years ago. It was speculated that the Pingtan people borrowed the idea of rock-cleaning from people in Jinmen. The method, although very simple, is quite effective, and for a long time Jinmen has had the best production of glueweed and Pingtan that of laver in China.

Two more seaweeds have also been cultivated for some years—in Dalian the *haidai*, or Japanese kelp (*Laminaria japonica*), which came from Japan in 1927, and in Qingdao the *qundaicai*, or wakame (*Undaria pinnatifida*), transplanted in the late 1930s from southern Korea. In both cases, cultivation was effected by simply throwing stones into the sublittoral seabeds during the spore-shedding season to provide the necessary substratum for the kelp spores to settle, develop, and grow. In all cases of early seaweed cultivation, natural substrates, principally stones and rocks, were employed for collecting the spores in nature.

The oyster was cultured by planting bamboo sticks or slabs of stone vertically in the lower tidal zones for spat-setting. Cultivation of the benthic razor clam (*Sinnovacula constricta*) was recorded in the city records of Chaozhou, eastern Guangdong Province, a few hundred years ago. A few other benthic mollusks, such as the ark shell (*Arca granosa*) and the clam (*Raditopes philippinarum*), were also under cultivation by merely scattering their natural "seed" on selected seabeds.

Another type of traditional mariculture in existence for a few hundred years may be called "inlet-cultivation." Small inlets with narrow mouths in sheltered coastal regions were selected and at some time in spring, when the fry of the

Table 1
Total mariculture and marine fishery production in China, 1950–1990

Year	Marine Fisheries Production (in 10³ metric tons)	Mariculture Production				
		Total (in 10³ metric tons)	As percentage of marine fisheries production	As percentage of total fisheries production	Cultivation area x 10³	
					mu	ha
1950	546	10	1.87	1.11	10	0.67
1955	1,656	107	6.46			
1960	1,870	121	6.47	4.0	152	10.1
1965	2,014	104	5.16			
1970	2,281	184	8.07	5.8		
1975	3,499	279	8.33			
1980	3,257	444	13.63	9.9	200.3	13.4
1983	3,617	545	15.07	10.0	280.1	18.7
1986	4,754	858	18.05	10.4	487.9	32.5
1987	5,480	1,100	20.07		553.9	36.9
1988	6,050	1,425	23.16		619.9	41.3
1989	6,640	1,576	23.80	13.7		
1990	7,130	1,620	22.70	13.1		

Source: Agricultural Economics Institute of the Chinese Academy of Social Sciences (1985) and Mr. L. Liang of the State Fisheries Administration.

Note: One mu is equivalent to 1/15 of a hectare (ha).

shrimp *Penaeus chinensis* and of the mullet *Liza haematocheila* were abundant in the water, were fenced off so that the young creatures were held in captivity until ready for harvest.

In summary, although more than 10 marine plants and animals were subjected to mariculture in old China, production was usually very low, amounting to only about 10,000 tons per annum, because of the low efficiency of the traditional methods, which had to depend on nature for their "seeds" and, in the case of sessile benthic organisms, their growth substrates.

Mariculture as a modern science is rather new in China. The difference between the old and new mariculture lies in the way the seeds are supplied. In old mariculture the seeds had to come from nature, whereas in new mariculture they come from artificial sources at the will of humankind. The seeds may be the spores of seaweeds, the larvae of mollusks and shrimp, or the fry of fish and are produced in confined places in nurseries; for the sessile benthic organisms the substrates are no longer the massive rocks and stones of nature but something much lighter and made specially for the organism's attachment.

A look at mariculture production in China during the period 1950–1990 shows that it had a very slow start, from a mere 10,000 tons in 1950, but had moved to 1,620,000 tons in 1990, a factor of 162 during four decades (see table 1).

The same data show that mariculture production has also accounted for an increasing percentage of the total marine fisheries production during the same period. In 1950, when new China was born, the 10,000 tons of mariculture production by traditional methods was only 1.87 percent of the total marine fisheries production, but by 1990 mariculture production accounted for 22.7 percent of total marine fisheries production. The cultivation area also jumped from 10,000 mu or 667 hectares in 1950 to 619,900 mu or 41,326 hectares in 1988.

Seaweed Mariculture

Japanese Kelp (*Laminaria japonica*)

Expansion of kelp production from a maximal pre-1950 annual output of 40.3 tons (dry) in 1949 to 6,253.5 tons in 1958 shows a 155-fold increase within a few years. The big increase in production was made possible by the introduction of artificial substrates in the construction of rafts for kelp growth. Before 1952, kelp cultivation was entirely by the method of throwing stones to the sea bottom or breaking submarine rocks to enhance the growth of kelp on natural substrate. The first successful experiment was made in 1952, when spores were collected on artificial substrates made of palm ropes in the form of floating rafts. About 10 tons of dry kelp were harvested, and this marked the first time in the history of mariculture that kelp was grown and harvested on artificial substrates in the form of rafts. The employment of the cultivation technique involving floating rafts made of artificial substrate is one of the keys to the success and development of the Chinese kelp phycoculture industry. When the floating raft method was first introduced, most mariculturists were quite skeptical of its merit and still favored the so-called sea-bottom cultivation on natural rocks and stones. Raft culture production began to lead in 1956, with 391.7 tons compared with 167.7 tons produced by sea-bottom culture (see table 2). In 1957 and 1958, the superiority of the raft culture shows even more distinctly. Since then, cultivation on natural rocks on the sea bottom has been abandoned as the principal method for commercial seaweed cultivation.

Success in the research on open-sea fertilizer application is another factor pushing kelp mariculture production forward. By means of porus clay bottles as tools for fertilizer application, commercial cultivation of the kelp was extended to the so-called outer sea regions of the Yellow Sea, originally too infertile for kelp cultivation. Since the early 1960s, the kelp industry of China has greatly expanded and individual kelp farms may each extend to several hectares, even tens of hectares. On the basis of the quick absorption of the fertilizer by the kelp and the effectiveness of intermittent application, periodic sprinkling of fertilizer solutions was employed, and eventually mechanical spraying of fertilizer solutions was developed and is now the standard method of fertilizer application in the Yellow Sea region.

The summer-sporeling low-temperature cultivation method of kelp, as originally devised, was to solve the weed problem,

Table 2
Natural substrate (rock) and artificial substrate (palm rope) production of Japanese kelp (*Laminaria japonica*) in China, 1946–1958 (in tons, dry weight)

Year	Production on natural substrate	Production on artificial substrate	Total	Artificial substate production as percentage of total production
1946	24.5			
1947	17.3			
1948	30.0			
1949	40.3			
1950	0.7			
1951	–			
1952	12.0	10.3	22.3	46.4
1953	86.5	28.2	114.7	24.5
1954	251.2	78.7	254.9	30.9
1955	321.8	206.0	527.8	39.0
1956	167.7	391.7	559.4	70.0
1957	580.8	1,439.3	2,020.1	71.3
1958	986.2	5,267.3	6,253.5	84.2

Source: Based on data in Tseng (Zeng 1984), original figures in fresh weight converted to dry weight by a factor of 1/6.

because by this method the kelp sporelings have a two-month advantage over the traditional method, thus avoiding the competition by weeds for space, light, and nutrients. The method has still more advantages. The two months' advantage increases production by 30 to 50 percent. The conditions for sporeling transplanting in late November to early December are much milder than those of the frigid winter in January and February. The method also saves cultivation materials, thus lowering the cost of production.

Southward extension of commercial cultivation of kelp is another innovation of China's kelp industry. The coastal waters of the East China Sea are very fertile with a high content of nitrogen nutrients. Studies on the relation of temperature to kelp growth showed that, although the optimal growth temperature is 5–10°C, some growth is still possible at 20°C and the kelp's reproductive sporangial sori can be formed at temperatures higher than 10°C. Thus, there is a sufficiently long period of low temperature in the East China Sea for the kelp to grow to commercial size and reproduce. This, in combination with the floating raft method of cultivation and the low-temperature culture of summer sporelings, has made the southward extension of kelp cultivation possible in the East China Sea region.

All the above successes in kelp studies resulted in substantial development of kelp production (see table 3). The development was possible also because of certain other factors. For instance, in 1971 there was an abrupt increase in production from 88,294 tons in 1970 to 133,970 tons in 1971—an increase of 52 percent in a single year—due to the large-scale employment of Japanese kelp as raw material for iodine extraction. On the other hand, there was a drop in production from 252,907 tons in 1980 to 219,518 tons in 1981, a decrease of 13 percent in one year; apparently there was an overproduction of kelp in 1980 and lower demand for kelp on the market. With the development of new kelp products for food purposes and increased demand for iodine, kelp production rose again and in 1989 a new peak production of 272,913 dry tons was reached.

Table 3
Mariculture production of Japanese kelp (*Laminaria japonica*) in China, 1959–1990
(in tons, dry weight)

Year	Production	Year	Production
1959	23,886	1977	222,211
1962	39,348	1980	252,907
1965	27,482	1983	231,296
1968	60,869	1986	203,437
1971	133,970	1989	272,913
1974	144,707	1990	244,000

Source: Zhu (1980) and production figures supplied by Mr. L. Liang of the State Fisheries Administration.

The Purple Laver (*Porphyra spp.*)

Another development of China's mariculture production is the modernization of *Porphyra* mariculture. Traditional cultivation by the rock-cleaning method for the last two centuries had helped in promoting *Porphyra* production to some extent. It was, however, only after solving a series of scientific problems that *Porphyra* mariculture became a firm industry.

First, the delineation of the *Porphyra* life history solved the "problem of the missing link" by identifying the spore that gives rise to the leafy stage. It was shown that the *Conchocelis* stage, when mature, forms conchospores that after liberation germinate to the leafy *Porphyra*. Studies on the biology of the *Conchocelis* also resulted in a better understanding of the best conditions for the growth of the vegetative stage and the production of the conchospores. This provided the basis for devising methods of mass culture and of intensive collection of spores by controlling the light intensity and light period as well as the culture temperature.

In 1963 the Ministry of Fisheries organized a national campaign for mariculture of *Porphyra haitanensis* in Fujian. There was a difference of opinion as to how the seedlings could be obtained, one group favoring natural seedlings and another favoring artificial seedlings. The latter thinking prevailed, and thereafter *Porphyra haitanensis* mariculture has followed the artificial method by first cultivating the *Conchocelis* stage in late spring and liberating and growing the laver conchospores in late autumn.

Experimental ecological studies of the leafy stage, especially the temperature factor and the exposure factor for intertidal species of *Porphyra*, resulted in devising a good method for cultivation. An improvement of the traditional Japanese pillar method of cultivation led to a new device known as the semi-floating method, which combines the strong points of the pillar and the floating methods and is especially useful for intertidal cultivation.

Modernization of *Porphyra* mariculture has been instrumental in the success and the development of the industry. Table 4 summarizes the development of the production of two common species of *Porphyra*, namely, the southern species

Table 4
Mariculture production of Purple Laver (*Porphyra haitenensis* and *P. yezoensis*) in China, 1971–1990 (in tons, dry weight)

Year	Production	Year	Production
1971	3,497	1980	7,212
1972	3,509	1981	7,029
1973	4,074	1982	6,815
1974	3,783	1983	9,987
1975	4,466	1984	12,374
1976	5,395	1986	13,586
1977	9,191	1988	15,576
1978	8,427	1989	9,255
1979	6,630	1990	8,800

Source: Zhu (1980) and production figures supplied by Mr. L. Liang of the State Fisheries Administration.

P. haitanensis and the northern species *P. yezoensis*, with a total 1988 production of 15,576 tons, equivalent to about 160,000 tons in wet weight. The production dropped in 1989 to 9,255 tons and in 1990 to 8,800 tons, but potential production has been successfully demonstrated.

Other Seaweeds

Cultivation of other seaweeds amounted to 11,500 tons in 1989 and 16,200 tons in 1990. *Undaria pinnatifida* has been cultivated by methods similar to those used for the Japanese kelp; the sporophyte can be raised under normal summer temperatures so no artificial cooling is needed. The dried product is principally exported to Japan. *Eucheuma gelatinae* has been cultivated by breaking the thallus into several pieces, each tied to a piece of dead coral, thrown into the subtidal region, and arranged in order by divers. The production is very limited, amounting to about 300 tons per annum. The Philippine carageenophyte *Kappaphycus alvarezii* has recently been introduced from the Philippines. Similar methods involving use of thallus fragments are applied to *Gelidium amansii*,

Gracilaria tenuistipitata var. *liui,* and *Hizikia fusiformis* cultivation. Propagation by means of spores is still under investigation. A near relative of the *Porphyra,* namely, *Bangia fuscopurpurea,* has also been cultivated by a similar method. The earliest cultivated species, *Gloiopeltis furcata,* presumably is still being cultivated by the primitive rock-cleaning method. Two microalgae, *Spirulina platensis,* an alga with especially high protein contents, and *Dunaliella salina,* an alga for extracting B-carotene, have been cultivated in recent years. Thus, there are 12 distinct marine algae now subjected to mariculture.

Mollusk Mariculture

Molluscan cultivation started with oysters in the provinces Fujian and Guangdong. The traditional method utilizes natural spats and stone slabs and bamboo sticks.

The Mussel (*Mytilus galloprovincialis*)

Mussel mariculture is very well developed in Europe and is well known for its enormous productivity. In China the mussel was originally studied in the 1950s as a fouling organism that needed to be eradicated. In the early 1970s, experimental mussel culture was initiated.

Studies on the biology of mussels in the Zhifu Bay of northern Shandong showed that the larvae occur from May extending to early July, which is also the season for harvesting kelp. The studies also showed that, if the kelp cords of the longline rafts were left in the water, the planktonic larvae of the mussels would readily attach themselves to the cords, thus effecting very efficient spat collecting. This method of collecting natural mussel spats is now widely accepted in mussel mariculture.

In places where natural spats are few or absent, artificial rearing of spats is necessary. By employing improved collectors in the rearing tank, providing larval food supply, and inhibiting bacterial growth by antibiotics in the rearing water, more than 10 million spat averaging 350–400 microns in length each can be steadily produced per m^3 of rearing tank volume.

Table 5
Mariculture production of the mussel in China, 1970–1990
(in tons, fresh weight)

Year	Production	Average Production per Hectare
1970	36,730	
1971	45,870	
1972	51,170	
1973	44,130	
1974	49,700	
1975	55,740	
1976	59,470	
1977	102,075	
1978	96,190	36.26
1979	65,990	54.38
1980	64,000	50.00
1981	95,500	67.62
1983	114,500	67.62
1986	210,657	
1988	430,000	120.30
1989	490,514	120.00
1990	495,000	117.90

In cultivation of mussels, the floating raft method is widely accepted. Mussels are cultured for six months to one year and harvested in spring or autumn. One longline raft can produce from 750 to 2,500 kilograms of fresh mussel per year, averaging 1,500 kilograms in Jiaozhou Bay. Table 5 shows mariculture production of the edible mussel in China during the period 1970–1990, from 36,730 tons fresh weight in 1970 to 495,000 tons in 1990. Production per hectare also increased from 36.26 tons in 1978 to 120.00 tons in 1989.

By 1977, the production area had reached 40,000 mu or 2,667 hectares, and products harvested reached 102,075 tons. Because of poor sales, the production area was gradually reduced until 1980 when it reached only 19,000 mu or 1,267 hectares, producing only 64,000 tons of the mussel. Because of the need for live feed for shrimp cultivation, the mussel was tried and found to be excellent. There was then a good demand for the *Mytilus,* and following growth of the shrimp cultivation industry, *Mytilus* mariculture bloomed again; beginning with 1986, its production (210,657 tons) surpassed that of *Laminaria* (203,437 tons), and *Mytilus* became the major mariculture product in China. In 1988, there were 60,000 mu under production, producing 430,000 tons of the product, averaging 8,020 kilograms per mu, whereas in 1983 production averaged only 4,507 kilograms per mu. In 1989, *Mytilus* production increased further to 490,514 tons, averaging 8,020 kilograms per mu or 120 tons per hectare. In 1990, there was a slight increase of production to 495,000 tons, averaging 7,857 kilograms per mu or 118 tons per hectare. With the further increase of shrimp production there is no doubt that more *Mytilus* will be needed.

The Scallops

So far, two species of scallops have been subjected to mariculture, namely, the local scallop *Chlamys farreri* and the introduced bay scallop *Argopecten irradians.* The former takes two years to mature; for the latter one year is sufficient. In 1980 the cultivation area was only 40 hectares, producing 54 tons, averaging 1.35 tons per hectare. In 1983 the cultivation area increased to 200 hectares producing 2,000 tons, averaging 10 tons per hectare. In 1988 the cultivation area increased further to 3,533 hectares, producing 120,000 tons and averaging 34.0 tons per hectare, a 25.2-fold productivity increase.

The local scallop has been subjected to cultivation since the 1970s. In 1974 experiments to stimulate production of its larvae were carried out in the laboratory. Partial drying, circulating seawater, and raising the culture temperature were adopted as cultural methods, and good results were obtained. In 1974, 68,800 spats 1 mm long were produced from 1 m^3 of culture volume. The spats, after attaching to hard substrate, are grown in cages of several layers. Commercial production of the local scallop started in the early 1980s.

The bay scallop was successfully introduced from the United States in December 1982, and mariculture was commercialized in 1985. In 1986, 23,686 tons of scallops, practically all of the local species, were produced; in 1989, 3,600 hectares of scallops were cultivated, producing about 129,491 tons, including about 50,000 tons of the bay scallop. Thus, in four short years the bay scallop's production, which started at practically nothing, rose to 50,000 tons. The mariculture workers warmly welcomed the bay scallop merely because its mariculture shortens the economic turnover from two years to one year, although the "gan-bai"—the introduced bay scallop—is not as good as the local scallop.

The Oysters

Mariculture of oysters by the traditional cultivation method, although primitive, was more advanced than traditional mariculture of other marine organisms because it employed bamboo as the attaching substrate. If the seeds had been artificially produced and collected, the method would have been equivalent to scientific mariculture of the present day. As advanced as it was in old China, mariculture of oysters remains practically the same in modern China, although in recent years spats have been artificially produced.

Three species of oysters are involved: *Crassostrea sp.* of Fujian and Zhejiang provinces on the East China Sea coast, *Crassostrea rivularis* of Guangdong and Guangxi provinces on the South China Sea, and *Crassostrea dalianwhanensis* in North China, although there is very little activity in oyster culture in North China. For quite some time the culture has been scientifically modern because of the annual production of spats. In 1980, the cultivation area was only 20,800 hectares, producing 27,000 tons, and even in 1983 the area was 24,867 hectares, producing only 36,000 tons. In 1988, however, the area was almost doubled—42,200 hectares—and the production was 74,000 tons. The unit production also increased from 1,425 kilograms to 1,755 kilograms per hectare.

Other Mollusks

The razor clam *Sinnovacula* has been cultivated traditionally by bringing spats from seed-production grounds and scattering them on cultivation grounds; recently attempts have been made to grow spats artificially. In 1990, 140,000 tons were produced. Clams (*Raditopes philippinarum*) and ark shells (*Tegillarca granosa*) have also been similarly produced.

Shrimp Mariculture

As mentioned above, shrimp cultivation in North China has a long history of a few hundred years under a kind of "inlet" cultivation called "gangyang." Annually in March, the shrimp begin to migrate northward from their winter quarters in deep waters in the southern Yellow Sea to their spawning grounds, mainly in the Bohai Sea, where they spawn in May. After arriving at the spawning grounds, they scatter in favorable places for spawning where the seawater temperature is about 15°C. The juveniles and postlarvae grow enclosed in inlets where their food supply consists of what is present there. Under these primitive circumstances the yield is very low and unstable, usually less than 5–10 kilograms per mu, or 75–150 kilograms per hectare.

In new China, efforts were made to improve the traditional method of inlet cultivation. Success was first achieved in 1960 by the Institute of Oceanology of the Chinese Academy of Sciences in Qingdao. Shrimp farming has been developing very slowly, however, because of little support and low profits. To better promote shrimp culture in China, the State Fisheries Administration (formerly the Ministry of Fisheries) organized a joint research project on shrimp fry rearing in the late 1970s. Optimal conditions for temperature, water quality management, and hatchery feed supply were intensively studied and techniques for industrial production of shrimp fry were developed in the early 1980s (Liu 1990.)

Pond culture is now the principal form of shrimp mariculture. The growout ponds are generally constructed by building embankments in the intertidal zone. During the widest tidal ranges, seawater flows in and drains out of the ponds through sluice gates. In some cases, pumping systems are used. The size of shrimp ponds in China generally varies from about 2 to 10 hectares (mostly 3–5 hectares) and 1.5 to 2.0 meters deep. Initial stocking density of juvenile shrimp in a growout pond is usually 150,000–300,000 per hectare, but recently a lower stocking density (no more than 75,000–150,000 fry per hectare) has been found preferable. Just as in the case of raising shrimp larvae, water quality management and adequate food supply are the important conditions for growth of the cultured shrimp (Liu 1983). Annual production of cultured shrimp in China jumped from 0.45 tons in 1978 to 185.9 tons in 1989, an increase of about 400 times (see table 6) (Liu 1990).

The most common species of shrimp under cultivation is the Chinese shrimp *Penaeaus chinensis* (formerly *P. orientalis*). A few other species, such as *P. penicullatus* and *P. merguiensis*, are also cultivated in South China.

In North China, pond-cultured shrimp are mainly fed with live marine or brackish-water invertebrates, such as the thin-shelled small bivalve mollusks *Aloides laevis, Masculus senhonsie,* and *Laternula navicula,* small gastropods (*Umbonium spp.*), the clam *Raditopes philippinarum,* the razor clam

Table 6
Annual production of cultured shrimp in China, 1978–1990

Year	Production (in 10^3 metric tons)	Total Area (in 10^3 ha)	Average Production (kilograms per ha)
1978	0.45	1.3	35
1979	1.3	7.3	178
1980	2.6	9.3	279
1981	3.6	13.7	263
1982	7.0	16.5	424
1983	8.9	20.3	438
1984	19.3	33.4	578
1985	40.7	59.7	682
1986	82.8	85.2	972
1987	153.3	131.1	1,169
1988	199.0	244.4	814
1989	185.9		
1990	185.0	143.9	1,285

Source: Based on figures of Liu (1990), Liu and Zhang (1991), and Mr. L. Liang of the State Fisheries Administration.

Sinnovacula constricta, and the mussel *Mytillus galloprovincialis.* The latter is cultivated and its production has recently jumped to about half a million tons because of the need for feed for the shrimp under cultivation.

Fish Mariculture

Fish mariculture in China, consisting of enclosing the fry of the fish *Liza haematocheila* together with the shrimp *Penaeus chinensis,* has a rather long history of a few hundred years. Production was very small, however, only about 75–150 kilograms per hectare. Beginning in 1958, experiments were initiated in artificial production of fish fry, and some success has been achieved in the mugil and, more recently, in the sea breams *Pagrosomus* and *Sparus,* the flatfish *Paradichthys,* and the grouper *Epinephelus.* In addition, natural fry of the Japanese eel have been caught and reared. In 1980, the area under cultivation was 16,733 hectares, but the production was very low, amounting to only 2,600 tons, averaging 155 kilograms per hectare. In 1984, the area employed had grown to 45,267 hectares and production amounted to 9,400 tons, averaging 208 kilograms per hectare. In 1989, the area cultivated decreased to 40,133 hectares but production amounted to 36,400 tons, averaging 810 kilograms per hectare (see table 7). The chief production province is Guangdong, which employed 34,933 hectares in fish mariculture production in 1989, producing 32,700 tons and occupying 90.0 percent of the entire Chinese cultivation area.

No single dominant species of fish produced a sufficient amount to be specially recognized. Some of the fish, such as *Paradicthys,* are under cultivation when young and, when they reach a certain size, are released to the natural environment to increase natural resources by ranching. Positive results have been obtained, but the work is still in an early stage. The fish were mostly cultivated in cages. In view of the great variety of fish for which the fry have been artificially grown, a fish mariculture era will ultimately come. Time will perfect its growth and development. Many fish are carnivorous and problems of large-scale feed production must be solved before we are able to feed them in captivity. But the time will

Table 7
Production of principal mariculture organisms in China, 1978–1990
(in 10^3 metric tons and as percentage of total mariculture and fishery production)

Name of Organism	Production in 10^3 metric tons					As Percentage of Total Production				
	1978	1983	1986	1989	1990	1978	1983	1986	1989	1990
Japanese Kelp (*Laminaria japonica*)	256.2	231.3	203.4	272.9	244.0	26.0	42.0	23.7	17.7	16.8
Purple Laver (*Porphyra spp.*)	4.4	10.0	13.6	9.3	8.8	1.0	1.8	1.6	0.6	0.5
Other Seaweeds				11.5	16.2					
Seaweeds Total			217.0	293.7	269.0					
Blue Mussel	96.2	114.5	210.7	490.5	495.0	21.4	21.0	24.6	31.7	30.5
Razor Clam (*Sinnovacula*)	47.0	89.0	82.8	138.6	140.0	10.5	16.3	14.5	9.0	8.6
Oyster (*Ostrea, Crassostrea*)	31.0	35.5	55.0	73.2	82.0	6.2	6.5	6.4	4.7	5.6
Scallops (*Chlamys, Argapecten*)		2.0	23.7	129.5	147.0			2.8	8.4	9.1
Clams (*Raditopes philippinarum*)			41.6	57.1				4.8	3.7	
Ark Shells (*Tegillarca granosa*)			24.2	38.3				2.8	2.5	
Mollusk Total			438.0	927.1	1,120.					
Shrimp (*Penaeus chinensis*)	.45	9.0	82.8	185.9	185.0	0.3	1.6	9.7	12.0	11.4
Other Crustaceans				4.1						
Crustacean Total				190.0	185.0					
Fish				36.4	43.0					
Total				36.4	43.0				2.4	2.6
Mariculture Production Total	450	545	858	1,576.	1,620.					

definitely come when fish mariculture will be the principal form of mariculture.

Overall Development of Mariculture on Coastal China

From Table 7 it may be seen that for mariculture in China molluscan production now leads strongly, totaling 1,120,000 tons in 1990. At the very beginning, the seaweeds, especially the Japanese kelp (*Laminaria japonica*), led the production. After 1980, however, the percentage of seaweed in mariculture production gradually dropped to 44.2 percent (in 1983) but still led by a small margin. The percentage of seaweed dropped still further to 16.6 percent in 1990, whereas the percentage of the mollusks in mariculture production had climbed to 43.8 percent by 1983 and 69.1 percent in 1990. We may therefore call the period 1983 and before the seaweed era of mariculture and the period beginning in 1986 the molluscan era of mariculture. Shrimp and fish cultivation is catching up. There is no question that mariculture of shrimp and fish will eventually dominate mariculture production, leading to a shrimp and fish era.

Judging from the development and evaluation of China's mariculture and also impressive efforts in other countries, there is equally little doubt that the industry is here to stay. It will grow and increasingly provide high-quality, protein-rich food for human populations.

Mariculture and Biodiversity

Returning to the issue of whether mariculture is detrimental or beneficial to the environment and biodiversity, I have personally observed that in the mariculture of shrimp, inadequate feeding has resulted in feed sinking to the bottoms of cultivation basins, causing serious pollution to the environment and death to the organisms in the vicinity. Some years ago during the summer, there was a serious red tide along the northern coast of Shangdong Province, followed by large-scale mortality of cultivated shrimp and other animals. Some believed that the red tide was caused by overfeeding in shrimp cultivation. On the other hand, I have also observed abundant growth of organisms, including such economic forms as the sea cucumber, underneath the *Laminaria* kelp rafts. Kelp and scallops intercropped on the same rafts grow very well. Intercropping of the seaweed *Gracilaria* and shrimp or fish also produces good growth for both the plants and the animals.

These facts help us to arrive at the following conclusions supporting the belief that mariculture is broadly beneficial to oceanic biodiversity: (1) Raft culture of kelp is helpful to many other organisms nearby; (2) intercropping of seaweeds and animals is helpful to both of them in their growth as well as growth of other kinds of lives; and (3) mariculture of animals, especially shrimp and fish, whether in tanks or cages, must be carefully managed so that the feed can be practically all consumed and the leftovers cleaned; excess nutrients will then not cause pollution to the environment and therefore not be detrimental to biodiversity.

We must admit that the Yellow Sea coast and even the East China Sea coast are rather poor places for seaweeds to grow. Temperature variation in a year is too great, from 1–2°C in the coldest winter to 28°C in the hottest summer. Only a small number of perennial marine plants can tolerate such great difference in temperature, and only annual plants with a certain tolerance of season can enjoy living there. The local seawater is very turbid, light penetration is very low, and tidal difference is great. Only a small number of seaweeds can grow in such circumstances. Finally, seaweeds reproduce by unicellular spores and generally distribute to other places with the help of ocean currents, which carry their spores from place to place. *Laminaria japonica* is a Japan Sea and Japanese coastal North Pacific cold temperate species; it should grow happily in places like the Dalian area of the North Yellow Sea, but it did not grow there until the 1920s. Yet for the past more than one thousand years, the Chinese people have been importing *haidai* (*Laminaria japonica*) for food from Japan and Korea. This is simply because there is no current flowing between the two places.

At present, the *haidai* not only grows well in Dalian but is also commercially cultivated in Qingdao and even in Dongshan, Fujian Province, just a short distance from the Tropic of

Cancer. This is possible because of the application of two innovations by Chinese scientists, raft culture and summer sporelings. The *Laminaria* was introduced accidentally, or perhaps naturally, from Hokkaido, Japan, to Dalian, China, about 39°N, and grew quite well. In 1950 it was further transplanted to Qingdao at about 36°N. For growth of kelp in autumn, winter, and spring, when seawater temperature is lower then 20°C, there is no difficulty. But in the summer, when seawater temperature exceeds 24–25°C, as it does nearer the tropics on coastal China, the *Laminaria* fronds become unhealthy and start to die. With rafts, *Laminaria* workers can lower the kelp ropes to deeper and cooler parts of the sea. The *Laminaria* ropes can also be raised when the seawater becomes extremely turbid. In Zhejiang Province, for instance, seawater is very turbid and light penetration very low at certain times of the year, and the kelp ropes have to be raised almost to sea level so that the kelp fronds may receive the necessary light. If the kelps were attached to immobile rocks, they would undoubtedly die because of lack of necessary light or high temperature.

Laminaria was transplanted to south of the Changjiang River because the seawater in Zhejiang and Fujian provinces is extremely rich in nutrients and has a sufficiently long low-temperature period for the plant to grow to commercial size. The kelp, however, cannot survive the summer because of the high seawater temperature, reaching 30°C. Even in Qingdao more than 90 percent of the oversummering kelp fronds in deeper water were lost because of seawater temperature reaching more than 27°C and because of the big waves. Fortunately, we have invented the summer sporelings low-temperature cultivation method by which there is no need for mature *Laminaria* to live through summer and only sporelings need to survive summer, quite happily, under artificially cooled seawater.

Thus, in spite of the very poor growing conditions for the *Laminaria* in China, we have succeeded in cultivating 300,000 tons dry, or 1,800,000 tons fresh, kelp in 1991. Man has finally conquered nature using science and technology.

Coming back to biodiversity, we believe it is the raft that encourages growth of not only the organisms under cultivation but also other organisms growing "naturally" on it or nearby. When we let the raft remain in the sea in the summer, we find many kinds of unintended organisms, including seaweeds, mollusks, and crustaceans, growing on it. The rafts, together with the growing kelps, provide not only places for growth of attaching organisms but also shelters for those growing underneath them. The presence of algae in ecosystems is very important because algae are primary producers. Coral reefs and mangroves are good examples of ecosystems having high productivity and biodiversity. In some places, cultivation of kelp needs application of fertilizers, which will encourage production of not only the kelps but also other seaweeds and, eventually, also the animals. Unlike farming on land, for which pasture and forests have to be destroyed to give places for the grains, farming in the sea can utilize places not previously employed in any activity. We can therefore conclude, in this sense, that proper mariculture of seaweeds would not cause any environmental hazard and would have good influence on biodiversity. We have, of course, created an artificial or modified environment and community.

Mariculture of animals, especially shrimp and fish, may cause some environmental hazard if it is not adequately managed, especially in the feeding process. Undigested feed will result in putrefaction and consequently be detrimental to the environment as well as biodiversity. Therefore, steps must be taken to see that leftover animal feeds are well disposed in order to protect the environment and biodiversity; the economic incentives to do so are obvious, as this also protects the mariculture crop.

Mariculture is of great antiquity in China; it is not always easy to distinguish the natural distribution of an organism from one modified by man long ago.

The issue of the impact of mariculture on biodiversity can perhaps best or most importantly be thought of in terms of the difference in impact between gathering food value from natural marine environments and populations and raising the same food value in artificially modified marine environments, or, alternatively, making the same comparison with respect to land-based agriculture. With care in the introduction of species to new regions and with proper culture technologies to minimize impacts on local environments, mariculture can offer many benefits to the broad preservation of oceanic biodiversity. This broad value comes simply from reduced fishing pressure on natural populations, which, in turn, also helps to preserve the broad genetic diversity of the natural populations. Coastal mariculture also offers societal incentives to avoid and reduce coastal pollution. Diverse communities of marine organisms can thrive in association with and in the vicinity of the cultured organisms and our modified environments; although not necessarily the same communities or organisms that would have lived in the absence of the artificial environments, they can be healthy and diverse, particularly in the case of seaweed culture.

Thus, on balance, many aspects of mariculture can be of benefit to the care and preservation of biodiversity.

References

Acknowledgment: This is contribution No. 78 from the EMBL and No. 2151 from the Institute of Oceanology, Chinese Academy of Sciences.

Agricultural Economics Institute of the Chinese Academy of Social Sciences, Bureau of Fisheries of the Ministry of Agriculture, Animal Husbandry and Fisheries, and Chinese Society of Fishery Economics. 1985. *Fisheries economics of China (1949–1983)*. 1,798 pp.

IOEP (Section of Experimental Phyco-ecology, Institute of Oceanology, Academia Sinica). 1976. All-artificial spore-collecting cultivation of tiaoban zicai (*Porphyra yezoensis* Ueda). *Scientia Sinica* 19:253.

Liu, Ruiyu (J.Y. Liu). 1983. Shrimp mariculture studies in China. In *Proceedings of the 1st International Biennial Conference of Warm Water Aquaculture*, 82–90. Hawaii.

———. 1990. Present status and future prospects for shrimp mariculture in China. *Proceedings of Asian-U.S. Workshop on Shrimp Culture,* 16–28. Honolulu.

Liu, Shilu, and Mingdi Zhang. 1991. Present status and future development of mariculture in China (in Chinese). *Modern Fisheries Information* 6, no. 1:5–10.

Tseng, C.K. 1933. *Gloiopeltis* and the other economic seaweeds of Amoy, China. *Lingnan Scientific Journal* 12, no. 1:43–63.

———. 1981a. Commercial cultivation (of seaweeds). In *Biology of seaweeds,* ed. C.S. Lobban and M.J. Wynne, 680–725. Oxford: Scientific Publications.

———. 1981b. Marine phycoculture in China. *Proceedings of the International Seaweed Symposium* 10:123–152.

———. 1982. Some remarks on the kelp cultivation industry of China. *Proceedings of the 1981 International Gas Research Conference,* 728–733.

———. 1989. Farming and ranching of the sea in China. In *The future of science in China and the Third World,* ed. A.M. Hassan, 92–106. Proc. 2nd Gen. Conf. Org. Third World Acad. Sci.

Tseng, C.K., and T.J. Chang. 1955. Studies on the life history of *Porphyra tenera* Kjellm. *Scientia Sinica* 4:375–398.

Tseng, C.K., T.G. Liu, B.Y. Jiang, Y.H. Zhang, and C.Y. Wu. 1963. Studies on the growth and development of *haidai* (*Laminaria japonica*) transplanted at the Chekiang coast. *Stud. Mar. Sinica* 3:102–118.

Tseng, C.K., K.Y. Sun, and C.Y. Wu. 1955a. On the cultivation of *haitai* (*Laminaria japonica* Aresch) by summering young sporophytes at low temperature. *Acta Bot. Sinica* 4:255–264.

———. 1955b. Studies on fertilizer application in the cultivation of *haitai* (*Laminaria japonica* Aresch). *Acta Bot. Sinica* 4:374–392.

Wang, Zichin. 1981. Studies on spat-rearing and experimental cultivation of *Chlamys farrer* (in Chinese). *Jour. Dalian Fish. Coll.,* no. 1 (1981): 1–12.

Zeng Chengkui (Tseng, C.K.). 1984. Phycological research in the development of the Chinese seaweed industry. *Hydrobioloia* 116/117:7–18.

Zhang, Fusui. 1984. Mussel culture in China. *Aquaculture* 39:1–10.

Zhang, Fusui et al. 1986. A report on the introduction, spat-rearing, and experimental culture of Bay Scallop, *Argopecten irradians* Lamarck (in Chinese with English abstract). *Oceanol, limnol, Sinica* 17, no. 5:367–374.

Zhao, Chuanyin. 1991. Some problems and solution, to current fisheries resources enhancement (in Chinese). *Modern Fisheries Information* 6, no. 2:1–8.

Zhu, Deshan. 1980. A brief introduction to the fisheries of China. FAO Fisheries Circular, no. 726, iii+31 pp. Food and Agriculture Organization of the United Nations.

6
Microbial Diversity
Rita R. Colwell and Russell Hill

A greater understanding of microbial diversity in the marine environment is vital to gaining an appreciation for systems processes and for the influence of microbial populations on total biological diversity in this environment. The recent application of molecular techniques to the study of marine microbial diversity has revealed even greater complexities in bacterial populations than were previously discerned. In addition, the importance of viruses in marine ecosystems has become apparent only in the last three years, and the biodiversity of virus populations is totally unexplored. The study of marine microbial diversity is thus in the throes of a profound revolution, which has very important implications for the understanding of biodiversity of all marine life.

Conventional methods of microbial taxonomy, based primarily on biochemical characterization, have elucidated relationships between many of the culturable marine bacteria, notable in the important group Vibrionaceae, work done at the University of Maryland and elsewhere. Molecular techniques have also proved valuable in taxonomic studies, where sequencing of 5S and 16S rRNA (significant sections in ribosomal RNA, or ribonucleic acid) has been used extensively in determining the relationships between important marine strains.

In marine samples, typically less than 1 percent of the bacteria that can be visualized microscopically are culturable by current techniques. Many bacteria in natural aquatic ecosystems are present in viable but nonculturable state (Grimes et al. 1986; Roszak and Colwell 1987); this means that the assessment of overall bacterial biodiversity by means of only the culturable portion of the population is totally inadequate.

To study the biodiversity of the nonculturable portion of the marine bacterial population, it is necessary to use molecular techniques such as gene probing and polymerase chain reaction (PCR) amplification of 16S rRNA genes (Giovannoni et al. 1990). The superiority of gene probing techniques over conventional culturing techniques for the detection of specific strains in environmental samples has been demonstrated in our laboratory (Knight et al. 1990, 1991). The study of genetic diversity of marine bacterial populations has been greatly advanced by sequence analysis of PCR-amplified 16S rRNA genes. This technique has revealed the presence of a novel microbial group in the Sargasso Sea supporting the view that microbial ecosystems contain novel uncultivated species (Giovannoni et al. 1990). This has been further indicated by the remarkable finding that archaebacteria are present in pelagic marine bacterial populations (Fuhrman 1992). In our laboratory, this technique is being used to examine the diversity of bacteria present in black smoker fluid from the unique hyperthermal vent habitat (Straube and Colwell 1991; work in progress). This technique has the potential to greatly advance our understanding of the bacterial diversity of all parts of the marine environment.

In addition to studying existing microbial biodiversity, it is important to gauge the effects that anthropogenic forces are exerting on this biodiversity. Pollution may be having major, undetected effects on microbial diversity in the marine environment. Long-term monitoring programs are necessary to detect changes of this type. A good example of the type of program required is the Long-term Ecosystem Observatory (LEO-2500) that has been established on the Continental Slope off the coast of New Jersey under the direction of Fred Grassle of the Institute of Marine and Coastal Sciences at Rutgers. This site, where water depths are approximately 2,500 meters, has been impacted by sewage sludge dumping since 1986, and it has been shown that this sewage is affecting the benthic environment (Hill et al. 1992a,b; Bothner et al. 1992). This site serves as a useful model of the impact that widespread eutrophication of the benthic environment may have on microbial diversity. Indications are that the sludge dumping has had profound effects on the benthic microbial ecology (Straube et. al. 1992; Takizawa et al. 1992b). As part of this study, our laboratory has prepared DNA (deoxyribonucleic acid) samples of the total microbial population at heavily impacted sites and at unimpacted reference sites. These samples were prepared by a rapid, efficient technique (Somerville et al. 1989) that can be readily applied to a large number of environmental samples. These DNA samples yield a "genetic snapshot" of the total microbial population at a particular time and are an ideal basis for an archival record of the biodiversity of the population. Similar archival DNA information is being obtained for microbial populations in the Chesapeake Bay and the Bahamas and forms the basis of a valuable resource for assessing future changes in microbial diversity. It should be noted that microbes are ideally suited to this type of analyses, which are not possible to the same extent with macrobiota, and studies of this type may therefore be crucial in obtaining early warning of changes in biodiversity in the oceanic realm.

The recent discovery that viruses are far more numerous in marine systems than was previously realized (Bergh et al. 1989; Proctor and Fuhrman 1990) has interesting implications for microbial diversity in the oceans. Viruses may be an important factor in controlling bacterial populations, and it has been suggested that viral infections may be a source of selective pressure contributing to the high bacterial biodiversity in ocean systems (Giovannoni et al. 1990). Work in our laboratory has indicated seasonal changes in virus populations in the Chesapeake Bay, suggesting that viruses may be an important trophic factor (Wommack et al. 1992), and preliminary results indicate that viruses may be affecting bacterial productivity. It has also been shown that viruses can affect phytoplankton primary production (Suttle et al. 1990). By lysing bacteria and phytoplankton, viruses may divert carbon away from larger bacteriovores and herbivores and consequently return carbon, which would otherwise be utilized at higher trophic levels, to oceanic dissolved organic carbon pools. Viruses may therefore be an important factor influencing global carbon budgets, which in turn have a major impact on climate change.

In addition to the influence that viruses may have on microbial biodiversity, the biodiversity of virus populations in the marine environment should be considered. We have developed a differential filtration procedure that results in bacterium-free virus concentrates (Hill et al. 1992c), suitable for DNA extraction. Development of this method would be suitable to obtain genetic snapshots of virus DNA from a particular ecosystem, which could then be compared to subsequent samples to assess changes in virus biodiversity.

An understanding of the microbial biodiversity of the marine environment can also yield very tangible direct benefits. The marine environment is proving to be a valuable source of novel bioactive compounds with antibacterial, antiviral, and anticancer properties. Isolates from free-living bacteria and bacteria that are symbionts of marine invertebrates are likely to be a good source of useful bioactive compounds. Marine sponges in particular, which contain diverse communities of bacteria, produce many classes of compounds that are unique to the marine environment (Faulkner 1986). The maintenance of biodiversity in the marine environment is essential to preserving a valuable source of bioactive compounds.

An example of the tangible benefits that may accrue from an understanding of marine microbial biodiversity is provided by recent studies of the filamentous bacteria, actinomycetes. These studies indicate that the biodiversity of marine actinomycetes is considerably different from that of terrestrial actinomycetes (Weyland and Helmke 1988; Jensen et al. 1991; Takizawa and Hill 1992; Takizawa et al. 1992a). Actinomycetes are exceptionally important in biotechnology: they are remarkably versatile metabolically and produce more than 70 percent of the antibiotics currently in medical use. Traditionally, terrestrial actinomycetes have been screened for novel compounds but marine actinomycetes have been little investigated, although they have been shown to be ubiquitous in marine sediments. Following on from earlier work in our laboratory (Walker and Colwell 1975), we have recently found (Takizawa et al. 1992a) that actinomycetes isolated from the Chesapeake Bay comprise a markedly different species distribution from those found in terrestrial samples. Marine actinomycetes are potentially a very important source of novel bioactive compounds, but a necessary prerequisite for the biotechnological exploitation of this group is information on their biodiversity in the marine environment.

In conclusion, the study of microbial biodiversity in oceanic realms is of vital importance to the understanding of systems processes in this environment and can provide an early warning of major anthropogenic perturbations in this ecosystem. The study of marine microbial biodiversity will have useful consequences for biotechnology. In addition, it is likely to yield completely new information about the complex microbial communities in the environment. Marine microbial biodiversity is best studied by a combination of the techniques of conventional microbial ecology and the powerful molecular techniques that are now available and that have already revolutionized our understanding of the microbial biodiversity of the oceans.

References

Bergh, Ø., K.Y. Børsheim, G. Bratbak, and M. Heldal. 1989. High abundance of viruses found in aquatic environments. *Nature* (London) 340:467–468.

Bothner, M.H., H. Takada, I.T. Knight, R.T. Hill, B. Butman, J.W. Farrington, R.R. Colwell, and J.F. Grassle. 1992. Indicators of sewage contamination in sediments beneath a deep-ocean dumpsite off New York. Submitted.

Faulkner, D.J. 1986. Marine natural products. *Nat. Prod. Rept.* 1:251–280.

Fuhrman, J.A., K. McCallum, and A.A. Davis. 1992. Novel major archaebacterial group from marine plankton. *Nature* (London) 356:148–150

Giovannoni, S.J., T.B. Britschgi, C.L Moyer, and K.G. Field. 1990. Genetic diversity in Sargasso Sea bacterioplankton. *Nature* (London) 345:60–63.

Grimes, D.J., R.W. Atwell, P.R. Brayton, L.M. Palmer, D.M. Rollins, D.B. Roszak, F.L. Singleton, M.L. Tamplin, and R.R. Colwell. 1986. The fate of enteric pathogenic bacteria in estuarine and marine environments. *Microbiol. Sci.* 3:324–329.

Hill, R.T., I.T. Knight, M. Anikis, W.L. Straube, and R.R. Colwell. 1992a. Benthic distribution of sludge indicated by *Clostridium perfringens* spores at a sewage disposal site off

the coast of New Jersey. American Geophysical Union Ocean Sciences Meeting, New Orleans, La.
Hill, R.T., I.T. Knight, M. Anikis, and R.R. Colwell. 1992b. Benthic distribution of sewage sludge indicated at a deep-ocean dumpsite. Submitted.
Hill, R.T., K.E. Wommack, and R.R. Colwell, 1992c. Bacterium-bacteriophage interactions in the Chesapeake Bay. 92nd General Meeting of the American Society for Microbiology, New Orleans, La.
Jensen, P.R., R. Dwight, and W. Fenical. 1991. Distribution of actinomycetes in near-shore tropical marine sediments. *Appl. Environ. Microbiol.* 57:1102–1108.
Knight, I.T., S. Schults, C.W. Kaspar, and R.R. Colwell. 1990. Direct detection of *Salmonella spp.* in estuaries by using DNA probe. *Appl. Environ. Microbiol.* 59:1059–1066.
Knight, I.T., J. DiRuggiero, and R.R. Colwell. 1991. Direct detection of enteropathogenic bacteria in estuarine water using nucleic acid probes. *Wat. Sci. Tech.* 24:261–266.
Proctor, L.M., and J.A. Fuhrman. 1990. Viral mortality of marine bacteria and cyanobacteria. *Nature* (London) 343:60–62.
Roszak, D.B., and R.R. Colwell. 1987. Survival strategies of bacteria in the natural environment. *Microbiological Reviews* 51:365–379.
Somerville, C.C., I.T. Knight, W.L. Straube, and R. R. Colwell. 1989. Simple, rapid method for direct isolation of nucleic acids from aquatic environments. *Appl. Environ. Microbiol.* 55:548–554.
Straube, W.L., and R.R. Colwell. 1991. 16S rRNA sequences from selected areas of black smoker hydrothermal vents of the Juan de Fuca Ridge. Second International Marine Biotechnology Conference, Baltimore, Md.
Straube, W.L., M. Takizawa, R.T. Hill, and R.R. Colwell. 1992. Response of near-bottom pelagic bacterial community of a deepwater sewage disposal site to deep-sea conditions. American Geophysical Union Ocean Sciences Meeting, New Orleans, La.
Suttle, C.A., A.M. Chan, and M.T. Cottrell. 1990. Infection of phytoplankton by viruses and reduction of primary productivity. *Nature* (London) 347:467–469.

Takizawa, M., and R.T. Hill. 1992. Isolation and ecological studies of actinomycetes in the Chesapeake Bay. 92nd General Meeting of the American Society for Microbiology, New Orleans, La.

Takizawa, M., R.T. Hill, and R.R. Colwell. 1992a. Isolation and diversity of actinomycetes in the Chesapeake Bay. Submitted.

Takizawa, M., W.L. Straube, R.T. Hill, and R.R. Colwell. 1992b. Near-bottom pelagic bacterial communities at a deepwater sewage disposal site. In preparation.

Walker J.D., and R.R. Colwell. 1975. Factors affecting enumeration and isolation of actinomycetes from Chesapeake Bay and southeastern Atlantic Ocean sediments. *Marine Biol.* 30:193–201.

Weyland, H., and E. Helmke. 1988. Actinomycetes in the marine environment. In *The biology of actinomycetes '88*, ed. Y. Okami, T. Beppu, and H. Ogawara, 294–299. Proceedings of the 7th International Symposium on the Biology of Actinomycetes. Toyko: Japan Scientific Society Press.

Wommack, K.E., R.T. Hill, M. Kessel, E. Russek-Cohen, and R.R. Colwell. 1992. Distribution of viruses in the Chesapeake Bay. *Appl. Environ. Microbiol.* 58:2965–2970.

About the Ocean Policy Institute

The Ocean Policy Institute was established in 1991 as an autonomous division of the Pacific Forum, CSIS's Honolulu-based policy research institute, with the mandate to address marine environmental, resource, and conservation matters. The Institute is dedicated to providing an international forum for research and analysis of ocean-related issues and endeavors to foster the stewardship and utilization of ocean resources for the mutual advantage of all nations. Its particular emphasis is the application of scientific and technological analyses and capabilities to economic, environmental, and policy issues. We envision the Institute becoming the premier center in the Asia-Pacific region for innovative research and collaboration on ocean policy problems. The Institute's Hawaii base allows it ideal access to develop a network of ocean-related institutions and organizations in Asian and Pacific Island nations and the U.S. mainland.

The Institute's director, Dr. Melvin N.A. Peterson, is a distinguished oceanographer, formerly of the Scripps Institution as well as the first chief scientist of the National Oceanic and Atmospheric Administration. The Institute's Advisory Council is chaired by Dr. Frederick Seitz, former president of the National Academy of Sciences and former president of Rockefeller University. The Council, composed of accomplished international leaders in matters of ocean affairs and policy, meets periodically to help set the overall agenda for the Institute's policy research and inquiry.

The Institute plans to conduct two research programs during 1993, both of which embrace issues of importance to the short- and long-term productive and wise use of oceanic resources. The first project, in the field of environmental remediation, will focus on an evaluated analysis of practical options available to respond to the issues of contaminated sediments in various harbors, lagoons, estuaries, and other underwater locales. The project will include analysis of several case studies in the Pacific region.

The second project is to embark on the second phase of a study in marine biodiversity, which takes into account the findings of a preliminary Phase I study that was concluded in the summer of 1992. The findings from Phase I indicated the need to develop a design and format for a comprehensive data base and information system for oceanic life that would serve as a teaching, research, and environmental management tool and that would allow oceanographic knowledge to be integrated into problem-solving in a variety of marine fields. The Institute plans to undertake a series of workshops throughout 1993 with relevant international experts to accomplish this task.

Advisory Council
Ocean Policy Institute of Pacific Forum/CSIS

Frederick Seitz, *chairman;* former president, Rockefeller University, and former president, National Academy of Sciences (New York, New York)

William A. Nierenberg, *vice chairman;* former director of Scripps Institution of Oceanography and former vice chancellor, Marine Sciences, University of California at San Diego (La Jolla, California)

Members

Robert Bauer, former chairman of the board and founder of Global Marine, Inc. (Whittier, California)

Kenneth F. Brown, chairman of the board, Mauna Lani Resort, Inc. (Honolulu, Hawaii)

Adm. William Crowe, former chairman of the Joint Chiefs of Staff and currently a CSIS counselor (Washington, D.C.)

Mufi Hannemann, director, Office of International Relations, Governor's Office, State of Hawaii (Honolulu, Hawaii)

Sir Anthony Laughton, former director of the Institute of Oceanographic Sciences of the National Environmental Research Council, United Kingdom (Chiddingford, Surrey, UK)

Charles Merdinger, former deputy director of Scripps Institution of Oceanography (Incline Village, Nevada)

John Norton Moore, director, Center for Oceans Law and Policy, University of Virginia (Charlottesville, Virginia)

Noriyuki Nasu, professor, University of the Air, Japan; professor emeritus and former director of the Ocean Research Institute, University of Tokyo; and present chairman of the Council of Ocean Development of Japan (Tokyo, Japan)

Barry Raleigh, dean of the School of Ocean and Earth Science and Technology at the University of Hawaii (Honolulu, Hawaii)

David A. Ross, senior scientist and former director, Marine Policy and Ocean Management Center, Woods Hole Oceanographic Institution (Woods Hole, Massachusetts)

Nainoa Thompson, traditional navigator, Hokule'a expeditions recreating Polynesian voyages (Honolulu, Hawaii)

—— **Selected Titles in the Significant Issues Series** ——

Volume XIV–1992

The Chemical Weapons Convention: Implementation Issues
    Brad Roberts (ed.) $9.95

Diversity of Oceanic Life: An Evaluative Review
    Melvin N.A. Peterson (ed.) $14.95

Lusophone Africa, Portugal, and the United States: Possibilities for More Effective Cooperation
    Kimberly A. Hamilton $6.95

EC 92 and Changing Global Investment Patterns: Implications for the U.S.-EC Relationship
    Cynthia Day Wallace and John M. Kline $9.95

Post-Communist Economic Revolutions: How Big a Bang? (Creating the Post-Communist Order Series)
    Anders Åslund $9.95

Investing in Security: Economic Aid for Noneconomic Purposes
    Stanton H. Burnett $7.95

Dictionary of Political Parties and Organizations in Russia
    Vladimir Pribylovskii $16.95

The Atlantic Alliance Transformed
    David M. Abshire, Richard R. Burt, and R. James Woolsey $14.95

The Church in Contemporary Mexico
    George W. Grayson $9.95

Strengthening the U.S.-Japan Partnership in the 1990s: Ensuring the Alliance in an Unsure World
    Richard L. Grant (ed.) $8.95

The Last Leninists: The Uncertain Future of Asia's Communist States (Creating the Post-Communist Order Series)
    Robert A. Scalapino $14.95

Conflict Resolution and Democratization in Panama: Implications for U.S. Policy
    Eva Loser (ed.) $9.95

The U.S.-Japan Economic Relationship in East and Southeast Asia: A Policy Framework for Asia-Pacific Economic Cooperation
    Kaoru Okuizumi, Kent E. Calder, and Gerrit W. Gong (eds.) $16.95

Volume XIII–1991

The Venezuelan-U.S. Petroleum Relationship: Past, Present, and Future
    G. Henry M. Schuler $6.95

Red Armies in Crisis (Creating the Post-Communist Order Series)
    Bruce D. Porter $14.95

Security and Economics in the Asia-Pacific Region
    Gerrit W. Gong and Richard L. Grant (eds.) $9.95

Nuclear Power: The Promise of New Technologies
    Charles K. Ebinger, John P. Banks, and Margaret S. Morgan $8.95

CSIS Books   1800 K Street, N.W. Suite 400   Washington, D.C. 20006
Telephone (202) 775-3119   Fax (202) 775-3199